Total Quality Safety Management

—An Introduction

EDWARD E. ADAMS

T
55
.A35
1995

ILLINOIS WESLEYAN UNIV. LIBRARIES
P.O. BOX 2899
BLOOMINGTON, IL 61702

AMERICAN SOCIETY OF SAFETY ENGINEERS

Total Quality Safety Management—An Introduction
Copyright © 1995 by Edward E. Adams

All rights reserved under International and Pan-American Copyright Conventions. Published in the United States by the American Society of Safety Engineers. Excerpts from *The Theory Behind the Fourteen Points* reprinted by permission of Process Management International. Excerpts from "Deployment Flow Charting" reprinted by permission of Dr. Myron Tribus.

In Special Appreciation

This note is to express my everlasting gratitude and indebtedness to my Irish friend and computer guru Denny Hogan. Without his constant guidance, teaching, and rescuing, these pages would never have seen the ink from the presses.

—Ed Adams

Managing Editor: Michael F. Burditt
Copy Editor: Nancy E. Kaminski
Text composition: William M. Johnson
Cover design: Carlisle Communications, Ltd.

Library of Congress Cataloging-in-Publication Data

Adams, Edward E.
    Total quality safety management : an introduction / Edward E. Adams
       p.   cm.
    Includes bibliographical references and index.
    ISBN 1-885581-03-3
    1. Industrial safety—Management.   2. Total quality management.
  I. Title.
T55.A35 1995   95-13971
658.4'08—dc20   CIP

Manufactured in the United States of America

8 7 6 5 4 3 2 1

To Geraldine
The Song in my life

Contents

Foreword	vii
Preface	ix
1 The Paradigm Shift	1
2 Things are the way they are because they got that way	9
3 The Principles of Variation	19
4 What Is an Accident?	25
5 The Tactics and Strategies of Safety Management	33
6 The Concepts of Energy and Barriers	37
7 The Strategies of Prevention by Safety Program Systems	53
8 The Safety Responsibilities of Executive Management	87
9 The Management Oversight Risk Tree Diagrams Restated	105
10 Management by Fact	127
11 Employee Involvement in Safety Program Improvements	157
12 TRANSFORM: To Change Completely or Essentially in Composition or Structure	169
Index	197

Foreword

The imperatives for undertaking the challenge of producing the following pages were three in number, each consisting of a comment from someone whose opinion I held in high regard:

Comment #1: *"You know, Daddy, if your company ran their plants right they wouldn't need your job."* (Comment from Number Two Son, now manager of a small plastic film plant for a company undergoing transformation to the Deming Management Method.)

Comment #2: *"Ed, just what is safety all about? I try to work at it, but I just don't understand what it is. What is it I am supposed to be doing?"* (A comment from a young, ambitious, and talented Mexican–American supervisor in a large food processing plant in Texas. I was stuck for an answer.)

Comment #3: *"They've got to simplify it. It is just too complicated to be practical."* (A comment from a senior, and highly-respected safety management executive regarding the Management Oversight Risk Tree.)

My life–long interest in management strategies was spawned by Professor Walter Bogart as an undergraduate at Middlebury (Vermont) College, and was tremendously nurtured by participation in Professor Peter Drucker's first class of just thirteen students at NYU's Graduate School of Business. This class, which introduced us to Professor Drucker's concepts of "management by objectives and self–control," was just prior to the publication of his book, *The Practice of Management*. Finally, in the closing years of a 30–year career in safety management in a variety of industries, I was introduced to the teachings of Dr. W. Edwards Deming, which climaxed that interest.

As with too many safety professionals, the failure of most executive managements to vigorously support accident and loss prevention activities within their organizations has been of continuing deep concern. This management weakness was of such magnitude that action at the federal level led to passage of the OSH Act, providing the profession with the new imperative of compliance, an imperative which has done little to advance the economic function of safety within the organization.

The so–called "Quality Revolution" for American management was initiated in 1980 with the telecast of "If Japan Can, Why Can't We?" a program which many credit with catalyzing the re–examination of American management practices and principles. Today the "revolution" is well established in countless organizations including those of government departments. However, there is little supporting evidence that our safety and loss prevention profession has fully grasped the significance or magnitude of the changed philosophy.

And so these pages are the personal efforts of one safety and management student to interpret in a meaningful and new way the significance and methods of what I have chosen to call "Total Quality Safety Management". The exercise is the application of the statistical method of the flow chart, the Management Oversight Risk Tree, and other statistical analytical techniques to the Deming Management Method, and includes an interpretation of the significance of Dr. Deming's Fourteen Points for management to safety and loss prevention.

Preface

Total Quality Safety Management is a major accomplishment by a safety manager with years of experience in applying the Total Quality Management principles of W. Edwards Deming to the real–life safety practices of a large national corporation. To use the author's phrasing, it is a "roadmap" for continuous system improvement using the Management Oversight Risk Tree, and as such it is an invaluable tool for safety professionals seeking to analyze and assess the structure and responsiveness of their companies to safety concerns.

Identification of MORT subsystems provides the basis for the analysis of system variations that can result in hazard related incidents, which are defined as unwanted flows or transfers of energy that produce harm instead of beneficial work. Working diagrams for the subsystems are provided and related to the *Mort User's Manual*. The application of statistical methods in problem identification and analysis is also thoroughly explored.

The presentation of the MORT–based Safety Program System (SPS) and Management Safety System (MSS) summarizes the distinction between work that is tactical, affecting the way tasks are done, and work that is strategic, affecting the way the organization is managed: work that requires the participation of all people, and work that only management can accomplish. These flow charts should serve as working documents which will aid safety professionals in identifying the individual root cause subsystems that are candidates for improvement.

Total Quality Management is focused on continuous process improvement. While the process is improved, waste is being eliminated: all wastes by everyone in an organization. To achieve this combination of benefits, implementation of TQM requires that traditional management methods undergo a transformation—a paradigm shift. Internal management decisions can no longer be determined based on an often shifting power–based focus, but on cooperation which will produce maximum benefit for the entire organization. In *Total Quality Safety Management* Ed Adams provides a detailed analysis of the type of organizational structure which can provide the basis for continuous safety improvement.

1

The Paradigm Shift

Total Quality Management is one label among many that is applied to the American version of what is often called "The Japanese System," a system of management originally introduced to the Japanese by an American, Dr. W. Edwards Deming.

Dr. Deming is the man who told the Japanese how to organize and manage their industries using the relentless daily pursuit of increased quality and productivity as the operational objective for every business process. The objective of continuous improvement was applied originally to manufacturing processes, but it eventually grew to include administrative and financial processes, sales, distribution, and personnel; in short, everything. One defining principal of the new system was the elimination of waste—the waste of materials, the waste of assets, and the waste of the time of people, all people, all levels. And he provided the technical tools: simple statistical charts to do the job.

The year was 1950: under conditions that were ideal for introducing total change, the new system revolutionized Japanese management thinking. What occurred has been called a "paradigm shift." A paradigm is a model of thinking that represents the ideas considered to be true and best, the theories and knowledge that direct policies and actions. A paradigm shift occurs when a new model or system of ideas appears that destroys the old model of what was considered true and best. Some have compared the magnitude of this paradigm shift in management thinking to the shift that occurred when it was established that the world was round, not flat; or to the shift that occurred when Sir Isaac Newton discovered that all the laws of nature, including gravity, could be expressed in mathematical terms. Each of these discoveries reduced all previous theories to the status of folklore.

Management's First Responsibility

The bedrock principle upon which Dr. Deming's approach rests is that the first responsibility of management is to stay in business and create jobs. The immediate corollary of this principle is that the long term success of the business depends on satisfying customers.

The commitment to growth by satisfying customers is the operating touchstone for all in the organization. Everyone has a customer and must know who that customer is. Customers are internal as well as external. Who receives your work? Whom must you satisfy?

Traditional management, on the other hand, views the concept of "growth–by–satisfying–customers" as a strategic option, one among many. The leading American schools of business, starting in the 1950s, espoused management based primarily on financial goals—measured in financial terms—as the proper route to organizational growth, treating the enterprise as an "investment portfolio." For many managers the customer came to be viewed as the purchaser of the company's stock.

Myron Tribus has pointed out that management by financial results is managing by "lagging indicators," and that "trying to run a company by looking at resulting financial figures is like trying to drive a car guided by the white line one sees in the rear view mirror. It doesn't work."[1]

Other enterprises, particularly those involved in consumer products, adopted "increased market share" as the strategic option for growth. The marketing department became the fair-haired people. Production activities were viewed as something of a necessary evil, a cost that, if they could, they would eliminate altogether. For many, quality control was viewed as the price they paid to avoid liability lawsuits and, more recently, even possible imprisonment of their executives.

The structure built upon the foundation of satisfying customers differs from traditional management in three fundamental ways:

1. Its definition and use of the quality objective for all operating decisions.
2. Its focus of management attention on the processes that produce the results.
3. Its use of prescribed managerial methods to find problems and solve them.

The Mandate of Total Quality Management

The mandate of Total Quality Management is for every employee urgently and enthusiastically to want to continuously improve the quality and productivity of his or her work. This involves:

- The workers working within the system;
- The manager's staff working on improving the system with the workers' help;
- Upper management providing the wherewithal to make the improvements, through concept and policy, budgets and training, staff assistance, and personal vigor and example to create the conviction and belief in all the people that it can be done.

The Quality Objective

In TQM the quality objectives are specific: The product or service that results from my work must be the most useful, most economical, and always satisfying to my customer.

This definition and guide to action applies to all, no matter how far he or she is from the ultimate consumer. Customers are defined as the people who depend upon the output of the process or system. All customers, whether external or internal, immediate or ultimate, are equally important and must be treated as such. All customers must receive what they have a right to expect.

The traditional management formula leaves quality objectives unspecified. Individual managers are free to define them on their own terms, i.e. "manufacture to specification," to "reasonable tolerances"; or "produce quantity and inspect out defective products," "salvage through rework," "adjust customer complaints"—and so on. Rarely are quality objectives specified for internal customers.

---

1. Myron Tribus, "The Quality Imperative," *The Bent of Tau Beta Pi* (Spring 1987).

"Keeping close to the customer" is an adage in sales management. With customer satisfaction as the foundation of management thinking, total quality management is *always* "close to the customer." Traditional management is only as close to the customer to the extent that individual managers decide to make the effort.

The Focus of Management Attention

Traditional management focuses on *producing results*. Total quality management focuses on *managing the processes that produce the results*. This change in focus yields a fundamental redefinition of the manager's functional role.

Traditional management loosely defines managers' roles in terms of "planning," "organizing," "implementing" and "controlling." Managers are free to pick and choose their methods, providing they produce the results that are demanded. It is considered sufficient that they know what is expected.

In quality management managers' functional roles are tightly defined: the manager's job is to work *on* the system, to improve it *with* the worker's help. This redefinition has two parts:

1. It establishes the primary managers' task.
2. It requires that performance of the task be done with the help of those immediately involved—the workers.

The redefinition introduces two critical changes. Not only do managers now have a different focus of attention, they also must approach their job differently. The elements of teamwork and consensus decision making are introduced into the system, replacing the authoritarian processes of traditional management.

The Managerial Methods

Traditional management has no special preference for managerial methods. In fact, the history of management is one of experimentation that approaches faddism. Managers are encouraged to use different methods in different situations.

In Total Quality Management the preferences in managerial methods are strong, and using selected statistical and behavioral methods is required. These are the "basic tools" of the new technology. Having most of the people trained and proficient in using these tools at their respective levels is essential. Their use in identifying, analyzing, and solving problems is mandatory. They are viewed as the only way to assure continuous improvement of work.

The Process of Performance Improvement

There is a third element to the matter of focus and methods: the process of performance improvement. What is the process? Where does it start? How is it kept going continually for continuous improvement? The answers lie in the Shewhart Cycle.[2]

---

2. W. Edwards Deming, *Out of the Crisis* (Cambridge, Mass: Massachusetts Institute of Technology, Center for Advanced Engineering Study, 1986): 88.

Walter Shewhart was a statistician at the Bell Telephone Laboratories and is the recognized father of statistical process control. It was he who defined the limits of variation and developed the Shewhart Cycle. Dr. Deming was closely associated with Shewhart for several years, and it was Shewhart's theories that became the basis of Deming's work. In Japan, the Shewhart Cycle is called the Deming Cycle, for the man who introduced it there. It has also sometimes been referred to as the PDSA Cycle; *Plan, Do, Study, Act*. It is the operating process for continuous improvement. (See Figure 1-1.)

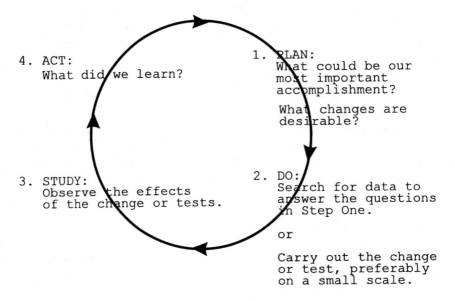

STEP 5: Repeat Step 1 with what was learned.
STEP 6: Repeat Step 2 and onward.

Figure 1-1. The Shewhart Cycle.

Step 1: The Planning Phase

Plan what you intend to do.
- Study the process to fully understand it and explore what changes might improve it.
- Identify the influencing factors. Organize the team to include people who have first hand experience and knowledge about them.
- Decide what data will be necessary. Does it already exist or will it be necessary to make a change and see what happens?
- Will tests have to be made?
- How will the data be used?

Do not proceed without a plan.

Step 2: Do It

Institute the plan developed in Step 1. Look for data on hand that will answer the questions raised, or carry out tests or make changes, preferably on a small scale.

Step 3: Study the Results

Record the effects of the changes or tests. What was learned? Look for side effects. Repeat the tests if necessary, perhaps in a different environment.

Step 4: Take Action

Adopt, alter or abort the plan. Do not be discouraged if some plans fail. If failure never occurs, the risks being taken are not big enough to pay off in big dividends.

Step 5: Back to the Drawing Board

Repeat Step 1 with the knowledge accumulated.

Step 6: Continue the Cycle

Repeat the steps in the Cycle.

Many will recognize the Shewhart Cycle. It was an essential point in production management training programs conducted during World War II. For many years, Step 3 had been called *Check*. In 1990, Dr. Deming changed *Check* to *Study*. He said *check* could mean to block, or pull up short, or rein in, and that's the last thing to be done at this stage.[3]

For some, the steps may seem to be little more than an alternative statement of the "planning, organizing, implementing, and controlling" functions of traditional management methods. That view misses the point—a point not missed by the Japanese and other practitioners of the new management methods.

The key to the Shewhart Cycle rests in Steps 5 and 6. In practice, the Japanese rigidly hold to following the reiterative nature of the process that is established in those steps. The objective is *continuous improvement of the process,* not solving a problem, or carrying out a project. The distinction is fundamental.

---

3. Lloyd Dobyns and Clare Crawford-Mason. *Quality or Else, The Revolution in Business* (Boston: Houghton Mifflin Company, 1991).

The Basic Questions

This discussion of the basic elements of Total Quality Management provides some indication of how the system works, and how the need to enable effective action is established. Each element poses a question that must be considered and for which answers must be formulated before action is contemplated.

1. Who is my customer?
2. What does my customer have the right to expect from me? What must I provide that will be most useful, most economical, and always satisfying to my customer?
3. What are the processes in my work that produce results? What is it that I must work on?
4. How can I apply the new tools, the statistical and behavioral methods, to improve the processes in my work that produce the results?

Applying the New Methods

Developed economies are based on two principal activities: producing "things," and providing services. The older and more fundamental of these activities is the production of things. This is where the statistical methods of management have been most widely applied and are most highly developed. (It is also what comes to mind most readily when the word "quality" is considered.)

The reason the new methods developed most rapidly in managing the production of things is simple. Making things results in physical items that can be counted (finite data) or measured (discrete data) and compared. The production processes are easily understood, and the customers who will use the things are easily identified. These characteristics make the foregoing questions easier to answer.

As you move from the activity of the production of things into the activity of providing services, the questions become more difficult. For some services, such as communications, transportation, banking and finance, and others, the answers are still not too difficult. Although "things" may no longer be directly measured or counted, they are tangible and easily perceived, and the "good" and "bad" characteristics of the service are identifiable. The processes involved in providing the services are easily understood, and the customers are not difficult to identify.

The greatest difficulties with clear answers to the questions arise when the service rendered is not tangible or easily perceived, the processes involved in providing the service are not easily understood, and there is confusion and diffusion on the question of customer identity. This difficulty is often encountered in measuring staff services provided to management. These characteristics describe the efforts of safety and health professionals.

The Shewhart Cycle: Step One

At this point we have arrived at Step 1 in the Shewhart Cycle, the need to study the process. If the process is not defined and understood, attempted changes will be misguided and could be disastrous.

Several years ago a young production supervisor asked me a very direct question: "Just what is safety all about? I try to work at it, but I really don't understand what it is."

I was embarrassed, for, at that time, I knew there was no way that I could sit down with him and in an hour or so help him with his question. I have never forgotten that young man or his question. But now, years later, I know one thing that happened that day. I had met a very important customer and I had failed him.

But the question "What is safety all about?" extends beyond that young supervisor. It bears directly on a matter that has been of deep continuing concern to safety practitioners throughout the evolution of our discipline: the matter of management support for our efforts. Scores of articles have been written on the subject, and it continues to be an agenda item at practically every safety conference or convention. The failure of safety professionals to gain universal acceptance of the role of accident prevention in traditional management practices has, finally, resulted in an evolving system of safety by government regulation. This development has resulted in a new safety program objective: maintaining compliance and thereby avoiding citations, fines, and even possible imprisonment. The development has done nothing to answer the question, "What is safety all about?"

If safety is to be successfully integrated into the new management methods, reexamination of the traditional model of thinking, the "paradigm of safety," is necessary. The explosion of safety by regulation makes it eminently clear that what has long been considered true and best has not served to the satisfaction of all.

The new management imperatives have redefined the understanding of the very word "quality". *Quality* is now understood to be *customer satisfaction*; it means *continuous improvement*; and it requires the *elimination of wastes* of every type—the waste of materials, the waste of assets, and the waste of the time of the people, all people, all levels.

The imperatives of quality management are demands that the traditional safety paradigm will be unable to meet. Superimposing the statistical and behavioral methods of the new paradigm onto traditional safety thinking will not lead to continuous improvement of our work. The need is to think our way out of the past and into the future with a new paradigm of "What is safety all about?"

2

Things are the way they are because they got that way

"Things are the way they are because they got that way. It helps to understand the history behind any problem before you attempt to change it."[1]

While the record of concern about safety, health, and accidents extends back six thousand years to the time the pyramids were built, the problems facing safety management today are of much more recent origin. The question "What is safety all about?" did not receive wide serious consideration until the turn of the century, one hundred years into what historian Arthur Toynbee called the Industrial Revolution.

At that time, the factory system was well established and expanding rapidly; cheap raw labor was flooding into the country; systems of distribution for products were developed and growing; new inventions were appearing at an unprecedented pace; the taming of the "wild" lands in the West was being completed; the Spanish-American War had been won; large corporations, banks, and insurance companies were flourishing; the gold standard was secure; orderly political systems were in place—all was well.

The Problem

Perhaps not quite all was well. There was one problem that had developed and was threatening to get much worse. That was the problem of worker injuries and deaths in the factories, mines, mills, and yards across the country.

The problem had four dimensions. First was the increasing number of workers, primarily immigrants from central, southern, and eastern Europe, being employed by these establishments. Heartland cities like Cleveland, Chicago, Detroit, Pittsburgh, and others were growing explosively. Small towns like Donora, Wheeling, Bethlehem, Peoria, and scores of others became industrial centers.

This increase in centers of population was related to the second dimension of the problem. That dimension was the increasing size of workplaces. While small businesses have always provided the most total employment, very large companies employing thousands were becoming more numerous. This directly affected public perception of the problem of worker casualties.

Worker casualties in small operations went virtually unnoticed by the general public, but in very large businesses, the injuries constituted a constant stream which attracted attention. And the perception of the problem was exacerbated by the large numbers of children employed in many industries.

---

1. Peter R. Scholtes, *The Team Handbook* (Madison, WI: Joiner Associates, 1988): 1-20.

The growing public outrage over the carnage that was occurring constituted the third dimension of the problem. The expression "butcher shop" became common. This outrage lead directly to the fourth and, from management's point of view, the most crucial dimension of the problem. That dimension was the erosion of the effectiveness of the legal defenses that employers had been able to use to control liability for worker casualties.

The Basic Law

The basic law governing employer liability for worker injuries rested in the common law that had developed over the preceding two or three hundred years. That had been an age where the production of "things" was carried out by individual craftsmen with a few helpers in small shops, and by workers who did the work in their own homes—the so-called "cottage" industry system. The law that developed gave the injured worker little chance for redress. Three doctrines were embodied in the law that strongly favored the employer:

1. *Fellow Servant Rule:* The employer was not liable for injury to an employee if that injury resulted from negligence of a fellow employee.

2. *Contributory Negligence:* The employer was not liable if the employee was injured due to his own negligence.

3. *Assumption of the Risk:* The employer was not liable because the employee took the job with full knowledge of the risks and hazards involved.

In the American system, derived as it is from the English system of jurisprudence, common law is administered by judges and peer juries. By the turn of the century, the public outrage at the inequities of strict interpretation of the provisions of the common law, as applied to job-related injuries, was resulting in decisions that no longer recognized these outmoded defenses. As a result, the number of claims being filed, the number of adverse decisions, and the amount of the damages being awarded were trending sharply upward. Management attributed these activities to an organized racket by unscrupulous lawyers and doctors.

To both industry management and their liability insurers the potential problems inherent in the situation were clear—a risk that would become uninsurable was rapidly developing. It was imperative that something be done to avert catastrophe.

The Solution: A New Principle of Insurance

The first successful legislative effort that addressed the matter of employer liability for accidents was made by Congress in response to a strong appeal from President Theodore Roosevelt. In 1908 Congress passed a worker's compensation act for federal employees. That limited act set the precedent for state laws to follow.

For manufacturers and their insurers, the arrangements and provisions of the various worker's compensation acts that developed proved to be the solution to the problem that was threatening catastrophe. Despite some resistance and adverse court decisions, based on what critics framed as a constitutional question of "taking property from the employer and giving it to his employee without due process of law," within a few years most of the industrialized states had adopted such laws.

The theory behind the new law for worker's compensation had its genesis in the social laws enacted in Germany during the latter part of the 19th century. There, the concept of entitlement to benefits without a qualifying test was established for the payment of worker's pensions—the first social security act. The challenge was to adapt this principal of governmental "insurance" to the requirements of private insurance contracting. The system that developed has become known as no-fault insurance, and it provided the insurance industry with a new and highly profitable product.

Compulsory worker's compensation was established as a no-fault insurance system. It made the employer liable for work-related deaths and injuries, regardless of whether there had been employee negligence. The three defenses of the common law were declared legally dead. The new law required employers to pay all the injured's medical expenses, to compensate them for any lost wages, and also compensate them for lost future wages if any permanent disability had been experienced. In the event of a fatality, the employer was required to compensate dependent survivors for their loss of the earnings of the deceased.

In return for the right to these incontestable benefits, the employees and their families gave up the right to bring suit for damages under the common law. The reasonableness of these provisions made the arrangement politically acceptable. In fact, the proposition was made to appear to be a magnanimous gesture on the part of the employers and their insurers. It was pointed out that the new law would encourage employers to undertake greater responsibility for preventing accidents in order to hold down their insurance premiums. This would be in addition to their moral responsibility to do so.

While the social benefits of the new arrangement were undeniably great, the economic benefits to the employers and their insurers were even greater. An uninsurable risk had been rendered insurable, and therefore budgetable.

How It Was Done

Examining the requirements for making an uninsurable risk insurable reveals the ingenuity of the new insuring contract.

The first requisite was to carefully prescribe exactly what was subject to coverage. The important qualifying phrase in the new laws was "injury arising out of and in the course of employment," and in many states the phrase "by accident" was included. That phrase indicated that there had to be an incident for the injury to be covered. A back injury, hernia, or hearing loss that had developed over time with no obvious incident was therefore not subject to coverage.

In time the "by accident" phrase was eliminated and special qualifying conditions were adopted to handle some of these "no incident" injuries. Characteristically, those qualifying conditions were highly restrictive. Other chronic illnesses, such as silicosis, black lung, asbestosis, etc., continued to be excluded. They were considered to be outside of the coverage and thus still subject to the rule of common law. Many years later these and other chronic, work-induced disabilities, redefined as occupational diseases and illnesses, have become a problem of momentous proportion..

The second requisite for making the risk insurable was to put a cap on the benefits that could be awarded. This was done by creating schedules of benefits that would be awarded injured employees. Three such schedules were needed: one to cover the loss of wages while the employee was temporarily disabled, one to compensate the employee for the loss of wages for any residual permanent disability, and finally there were death benefits to be paid to dependent survivors for the loss of the worker's earnings if a fatality had resulted.

The third requisite was to place the administration of the entire system in the hands of a commission, which was responsible for administering all phases of the act.

These responsibilities included administering appeal procedures and hearings in the event of disagreements, adjusting benefit schedules for inflation, recommending and implementing any changes in the basic act deemed necessary, and so on. The objective was to establish a system that was independent of the civil court system, enabling the injured to receive benefits due without having to resort to a lawsuit. This eliminated jury trials and also, it was felt, the shyster lawyer and quack doctor.

Finally, it must be noted that these were state laws. Historically, industry and its insurers had experienced little difficulty in influencing state legislators to respond favorably to their interests. Industry lobbyists worked diligently to limit the benefits and to restrict the number of claimants that would be eligible to collect them. In an article entitled, "Annals of Law, The Asbestos Industry on Trial," published in the June 10, 1985 issue of *The New Yorker*, author Paul Brodeur offered this comment on the compensation laws: "Workmen's Compensation proved to be a boon for employers: they were allowed to compensate for a worker's loss of life and limb at bargain prices during a time when the injury-and-death rate had soared in coal mines, steel mills, textile factories and other corners of the nation's workplace."[2]

Compulsory Worker's Compensation proved to be a boon to the insurance industry. It provided the industry with a brand new product, one that promised, based on the regulation of losses, to be very lucrative, with the potential for losses being regulated as they were. A few states opted to establish and administer their own insuring funds. However, most states elected to open the market to the commercial insurance companies. A very few chose to use both plans. Whichever plan was adopted, the programs that developed were strikingly similar.

Risk, Insurance, and Safety

Risk bearing is an everyday fact of life that arises out of natural or human-related forces. Risk is borne involuntarily, and it can only cause loss; it offers no hope of gain to anyone. The best that can happen is that things remain unchanged. Risk incidents that arise out of the forces of nature, such as floods, earthquakes, violent storms, etc. are often referred to as "acts of God". Risk incidents that arise out of human-related forces are generally referred to as accidents.

Risk, Uncertainty and Profit[3]

In the economics of the firm, there are two basic sets of circumstances out of which losses can arise—the circumstances of uncertainty and the circumstances of risk. The circumstances of uncertainty result from a conscious management decision to bear the uncertainty in the pursuit of profit, recognizing however that such decisions always involve the possibility of loss. The classic example of uncertainty bearing in the pursuit of profit is the introduction of a new product. The most prominent characteristic of uncertainty is the complete inability to forecast what will happen, regardless of the amount of past experience.

---

2. Paul Brodeur, "Annals of Law, The Asbestos Industry on Trial." *The New Yorker*, (June 10, 1985): 63.
3. This discussion is based upon the writings of the American economist Frank Knight in his book *Risk, Uncertainty and Profit* (Chicago: University of Chicago, 1971).

A second characteristic of uncertainty is that the greater the uncertainties, the higher the reward for success. Uncertainty bearing is the business of business. It is the true source of profit. It is the offensive, where the fun is, the high-visibility activity, where the heroes are made. The psyche of the entrepreneurial business manager is geared to uncertainty bearing, to taking the calculated chance, with an eye ever on the jackpot.

Risk bearing, out of which only losses can arise, differs radically from uncertainty bearing. The prevention of losses due to involuntarily-borne risks is a business's defense. The best that can happen is that you don't lose. It is dull, unromantic, with low visibility and no heroes; it is not much fun to even have to think about. It does not stir entrepreneurial juices.

The Management of Risk

The outstanding characteristic of risk is its predictability. A study of past experience will indicate the future. Where, when, and to whom an incident will occur cannot be predicted, but how often can be, with accuracy. It is this predictability that renders risk insurable. If enough risk bearers pool their funds, there will be sufficient funds available to cover the losses of the few in the group who experience the incidents. Risk can be reduced to certainty by pooling experience; uncertainty cannot, and is therefore uninsurable. However, for a risk to be insurable, it must be clearly defined. That is the purpose of the insurance contract, and worker's compensation acts are, in fact, insurance contracts.

In addition to being insurable, risk is subject to three other forms of management. All four risk management techniques involve expense. Management's problem is to select and apply the technique which handles risk most effectively for the expense involved.

The most elemental technique of risk management is to do nothing. For minor risks, where the loss potential is small and remote, this may be the valid answer. It could well cost more to do something than to just let it happen once in a while and pay the loss when it does.

However, risk situations vary all the way from minor to intolerable. Certainly for intolerable risks, such as those that threaten human or financial survivability, the do-nothing approach must be judged unsatisfactory. It is important, however, to recognize that doing nothing is a legitimate approach under certain circumstances.

The next two techniques of managing risk are closely related. They constitute the activities that are generally referred to as safety, or loss prevention. These are the techniques of risk elimination and reducing exposure to risk. These are the tasks that are of concern to safety and health practitioners. These concerns are not of modern origin. A brief account of the history of safety efforts can be found in the *Accident Prevention Manual for Industrial Operations*, published by the National Safety Council.[4]

In the twentieth century, a number of companies recognized that much more could be done to reduce and eliminate risk, even prior to the establishment of worker's compensation insurance. With the advent of Worker's Compensation, it was expected that these efforts would increase dramatically, which in fact they did. This occurred primarily in the heavier industries operated by very large companies, where the frequency, severity and visibility of accidents was high. For these large companies,

---

4. *Accident Prevention Manual for Industrial Operations, Administration and Programs.* 9th ed. (Chicago: National Safety Council, 1971): 2.

the advent of Worker's Compensation, with premiums related to the cost experience of their own accidents, the economic, as well as the moral imperative for effective safety programs was clear. However, this response was not universal.

Many companies did not have (or management felt they did not have) the resources necessary to provide safer equipment and working conditions. The funds necessary could be substantial, and identifiable returns on investment would be hard to come by. Opportunities for greater returns lay elsewhere. Since the losses were now insured and budgetable, the costs of injuries were viewed as under adequate control. There would be no surprises; the potential for catastrophic financial loss had been eliminated. The premiums were affordable and part of the cost of doing business.

The insurance carriers viewed this lack of management support with alarm, for it threatened the profitability of their worker's compensation insurance. Their conclusion was that neither management nor the workers understood what safety was all about. In the late 1920s they sought to rectify the problem.

Marketing the Product

In his seminars on "Quality, Productivity, and Competitive Position," Dr. Deming makes the point that we can learn only by the use of theory. "Without theory there is no knowledge," is his statement.[5] This statement reflects the actions the insurance industry took to solve the problem of management support for safety and injury prevention. In order to develop greater understanding of what safety was all about, a theory of accident causation was created. Not surprisingly, the theory that developed was based on the management theory of that time, the theory of "scientific management."

Scientific management was a product of the emerging discipline of industrial engineering. Frederick W. Taylor and Drs. Frank and Lillian Gilbreath were the foremost leaders in the development of the new thinking. "Technocracy" was the title applied, and its followers were called "technocrats."

The theory of scientific management is that human performance can be defined and controlled through work standards and rules. Using time and motion studies, jobs are broken down into simple, separate steps to be performed over and over again without deviation by different workers. Minimizing complexity maximizes efficiency. Under this system, workers are told just what to do and how to do it. Any change they make on the orders given them would prove fatal to their success. As a corollary, it was just as bad to overproduce as it was to underproduce in this system.

The theory of accident causation and the body of thinking—the paradigm—of the safety profession was molded by the industrial engineering-scientific management school. Its godfather, H. Waldo Heinrich, was a contemporary of Taylor and the Gilbreaths and was himself an industrial engineer, employed at the time in the engineering and loss-control division of The Traveler's Insurance Company. Heinrich's book, *Industrial Accident Prevention*, first appeared in 1931.[6] His philosophy and principles established the foundation of the new "safety engineering" discipline.

The theory and causation model that Heinrich developed is familiar to all. The theory was based on a set of propositions that Heinrich correctly labeled "axioms". *Webster's Third New International Dictionary* defines an axiom as "a proposition, principle, rule or maxim that has found general acceptance, or is thought to be worthy

---

5. Discussion based on W. Edwards Deming's book, *Out of the Crisis* (Cambridge, MA: MIT Center for Advanced Engineering Studies, 1986).
6. H. Waldo Heinrich, *Industrial Accident Prevention*, 1st ed. (New York: McGraw Hill, 1931).

thereof whether by virtue of a claim to intrinsic merit, or on the basis of an appeal to self evidence." In subsequent practice this definition was either overlooked, or disregarded, and the "axioms" were accorded the authority of scientific statements.

The first axiom was the theory of accident causation. Heinrich considered the occurrence of an injury to be the natural culmination of a series of events or circumstances which invariably occurred in a fixed and logical order, one dependent on another.[7]

The demonstration model was based on the simple chain reaction of falling dominoes. Five dominoes were used, each labeled as an event or cause factor. Starting with the result, the fifth domino was labeled INJURY. The preceding domino, the one that would cause the fifth to fall, was labeled ACCIDENT. The third domino, the one that would cause ACCIDENT to fall was given a double title, since there were two things involved. This established that accidents were caused either by something that someone did, (or failed to do), or by some condition that existed, or by a combination of the two. The third domino was labeled UNSAFE ACT/UNSAFE CONDITION.

At this point, the theory that human performance could be controlled through work standards and rules became dominant. It was a fact, observable by all, that practically every time someone was injured, the injury would not have occurred if someone only had not done something, or had done it differently. While it was recognized that mechanical or physical hazards might also be involved, the primary immediate cause was overwhelmingly most often the UNSAFE ACT.

Thus it was that the underlying cause was attributed to FAULTS OF PERSONS, the title given to the second domino. From there it followed that the problem had its roots in the ANCESTRY and SOCIAL ENVIRONMENT of the persons involved, the title given to the originating first domino.

In the fixed and logical order of the falling dominoes, removal of even one of the factors in the chain would not allow the injury to occur. In the prevention of injuries the target's bull's eye was clear! It was the center of the sequence, the UNSAFE ACT/UNSAFE CONDITION that had to be removed from the chain of events. And, by the observation and experience of all, the major problem was the actions of people.

The Folklore of Safety

With the theory of accident causation established, the next step was to build a body of knowledge that would serve as the paradigm for safety programs and activities. This was accomplished with the creation of Heinrich's axioms. The single most influential of the enabling axioms established the relationship between unsafe acts and unsafe conditions. The axiom asserted that 88% of all injuries were caused by unsafe acts, 10% by unsafe conditions and the remaining 2% were unpreventable.

The influence of this axiom cannot be over exaggerated. The belief that the overwhelming majority of accidents and work injuries are due to unsafe acts remains the basic conviction of most managers and supervisors, most of the general public, and even very large segments of the safety profession to this day. This is the legal theory of contributory negligence. Even though the compensation acts had declared the principal of contributory negligence legally dead, it remains alive and well in the folklore of safety.

---

7. This discussion is based on: H. Waldo Heinrich, *Industrial Accident Prevention, A Scientific Approach*, 3rd ed. (New York: McGraw-Hill, 1950).

Another axiom asserted that, on average, for every disabling injury caused by an unsafe act, there had been 29 minor injuries and over 300 narrow escapes from that same action. This created the "1-29-300 Ratio," often depicted as a segmented triangle.

This axiom led directly to the next axiom, where it was established that the severity of an injury is largely fortuitous. Luck alone had prevented it from being more serious. Therefore, every injury, regardless of its severity, must be investigated with equal thoroughness.

The final primary axiom stated that the humanitarian incentive for preventing injuries was supplemented by two powerful economic factors: first, that a safe business is efficient and productive, and second, that the direct cost of injuries to employers for compensation benefits and medical treatments is but one-fifth of the total cost the employer must pay. (The term *direct cost* was later changed to *uninsured cost* since they were direct only to the insurance company.) This relationship was often depicted as an iceberg, with the insured costs of medical treatments and disability benefits shown in the above-water portion of the iceberg, and the uninsured costs shown in the mass below the waterline.

Over the years, substantial attention has been directed to these uninsured costs, with the thought that they should be a powerful motivation for management to support safety programs. The most widely-recognized approach to the analysis of these costs is that developed by Grimaldi and Simonds.[8] They enumerated ten different categories of uninsured costs that could be identified and measured. They are:

1. The cost of wages paid for the working time lost by workers who were not injured.
2. The net cost to repair, replace, or straighten up material or equipment that was damaged in an accident.
3. The cost of wages paid for the working time lost by the injured workers, other than worker's compensation payments.
4. The extra cost due to overtime work made necessary by an accident.
5. The cost of wages paid supervisors while their time is spent on activities made necessary by the accident.
6. The wage cost due to the injured workers' decreased output after they return to work.
7. New workers' cost-of-learning period.
8. The uninsured medical costs borne by the company.
9. The cost of the time spent by higher-level supervisors and clerical workers on investigations or processing compensation application forms.
10. Miscellaneous unusual costs. (Demurrage, abnormal spoilage, loss of profit on lost orders, etc.)

With perhaps rare exceptions, the analysis of uninsured costs has not been effective in motivating management. The reasons this is true are not hard to find.

Not long before Heinrich developed his theories, a seemingly unrelated development had occurred on another management front, the accounting department. Here, the old "debit-credit" double entry bookkeeping system could no longer handle the complex task of factory accounting. Enter the accountants with the idea of standard cost accounting.

---

8. John V. Grimaldi and Rollin H. Simonds, *Safety Management*, 5th ed. (Boston: Richard D. Irwin, Inc., 1989): 214-19.

Based on a review of cost history, and using a series of assumptions, an effective system to handle the problem was developed. Unfortunately, these cost histories and assumptions included many elements that were purely and simply wastes and inefficiencies. Principal among these were the uninsured losses arising out of accidents of all types. These wastes were built into the cost standards; again, they were budgetable, with no surprises, a part of the cost of doing business, and assigned to obscurity in a system that could afford them. For safety, these two developments—the insurability of injury costs and the budgetability of accident wastes—created serious problems.

For the majority of managers, safety and accident prevention were not viewed as a profitable investment of limited resources. These managers envisioned no significant economic role for safety in management. With the cost of injuries handled by budgetable insurance and the uninsured cost of accidents of all types built into the standard costs, the only appeal left for safety was the moral one. The role of safety was not economic, but humanitarian. In management's eyes safety practitioners were viewed as missionaries.

Heinrich's book, *Industrial Accident Prevention, a Scientific Approach*, was first published in 1931. Since that time there has been considerable revision of many of the original concepts, some by Heinrich, many by others. However, many elements of the foundation that Heinrich laid down remain alive and influential after sixty years.

The Unspoken Assumption

The publication of Heinrich's book, and its subsequent promotion by the insurance industry did, however, result in the establishment of safety programs in scores of plants and industries where none had previously existed. Naturally, these safety programs were put in place in traditionally managed organizations. Traditional management is based on the hierarchical command-decisionmaking process derived from the sovereign right of kings, the military, much of organized religion, and other management organizations where the "right" idea is the one favored by the most powerful voice present. It is a system where dissenting ideas are not resolved, simply overruled. And, it is a system where decisions are strongly influenced by opinion, experience, impressions, intuition, secret agendas and hopes and wishes.

Myron Tribus has pointed out that the generic organization chart has a very long history.[9] In his workbook, *Deployment Flow Charting*, he makes a biblical reference to Exodus 18:21: "—and place over them to be rulers of thousands and rulers of hundreds, and rulers of fifties and rulers of ten."

This observation is followed with a diagram developed by the MANS (*Ma*nagement, *N*ew *S*tyle) organization in The Netherlands. The diagram is a traditional organization chart peopled with human figures—only the people at the bottom of the chart have no heads! This is the unspoken assumption of traditional management, and of the paradigm of traditional safety.

---

9. Myron Tribus, *Deployment Flow Charting*, vols I and II (Los Angeles: Quality & Productivity, Inc., 1989) (a workbook that accompanies Dr. Tribus's videotapes of the same title).

3

The Principles of Variation

A view from a different point at the same time

The Scientific Approach

Heinrich's problem centered on undesirable variations in workplace operations that were resulting in work injuries. By observation it had been determined that the great majority of injuries, some 85% or more, were the result of things that workers did—unsafe acts. The remainder were attributed primarily to the existence of improper mechanical or physical conditions—unsafe conditions.

The target was clear! Every effort must be made to remove the unsafe acts and unsafe conditions from the operation. The falling dominoes graphically demonstrated the causes and effects in the accident process. The objective was the prevention of injuries. The program required action on three fronts:

- *Engineering*, to improve the mechanical and physical conditions
- *Education*, to teach workers the safe way to do the job
- *Enforcement*, the use of authority to ensure that the instructions were being followed without deviation.

The program was based on controlling the undesirable results: injuries that were occurring in work processes.

During this period, the late 1920s, others were at work on a different version of the problems of variations in ongoing operations. On the other side of New York City, Walter Shewhart was studying variations of a different sort. Shewhart's assignment was the problems being projected for the future of telephone service.

Many problems had been identified, but three were particularly important to his work:

1. There was explosive growth in demand for telephone service in the immediate offing.
2. The manual switchboards in use at that time would not be able to handle the volume of traffic anticipated. The development of automatic switching was imperative.
3. To ensure acceptance, the new service would have to provide unvarying reliability to customers.

Shewhart approached his problems as a scientist and mathematician, employing newly-developing statistical methods in his work. Those new methods came largely from agricultural research projects on cause-and-effect relationships on crop yields, carried out in Great Britain under the direction of the British statistician R.A. Fisher. Shewhart transformed Fisher's methods into a quality control discipline for factories. This in turn later became the basis of the work of Drs. Deming and Joseph M. Juran on statistical quality control.

What are Statistics?

The word *statistics* usually brings to mind numerical data. For example, economic statistics are data on employment, production, prices, sales, earnings, and similar activities. Social statistics deal with population, education, welfare, crime, delinquency, drug addiction and the like. Statistics in sports are used to measure individual and team performance. An outstanding collection of statistics is in the *Statistical Abstract of the United States*, published by the U.S. Department of Commerce, Bureau of the Census.

However, in addition to numerical data, the word *statistics* has another meaning. It refers to the methods for gathering, analyzing, and drawing conclusions from factual information. It is a body of theory and methods which helps us make wise decisions under conditions of uncertainty.[1] Some of the most useful statistical tools involved in total quality management do not involve any mathematical data at all. In fact, the basis of Shewhart's sophisticated analytical charts is a nonmathematical statistical chart, the run chart.

The Run Chart

The run chart is one of the most simple statistical tools. It involves plotting data over a period of time. In his work, Shewhart might have plotted the number of inputs that were faulted by the experimental switchgear in a series of trials. This was enumerative data, data attained by simple counting. With the results of a statistically significant number of trial runs in hand, the result of each trial was plotted on a chart. This produced the basic run chart, a plot of test results over a series of tests, as shown in Figure 3-1. No mathematical calculations were involved.

Shewhart's next step was analytical, determining the average number of faults (variations) for the entire group of trials. This value was charted as a line around

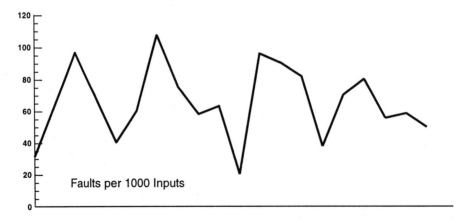

Figure 3-1. The basic run chart.

---

1. These definitions are taken from a book authored by Chris A. Theodore, *Managerial Statistics: A Unified Approach* (Boston: Kent Publishing Company, 1982).

which the individual plots would be located. This created a more refined picture of the system as it had operated over the series of tests. It differentiates between desirable and undesirable variations, as shown in Figure 3-2.

Since Shewhart's objective was improvement of the system, the picture made the targets clear. If the factor which caused a measured condition above the average for the system could be identified, and either eliminated or minimized, the average would be nudged lower, indicating improved performance. Likewise, if whatever caused a point to fall below the average could be identified and repeated, the average would be pulled lower, again improving performance. As these actions were taken, the spread between the points above the line and those below the line would narrow, indicating greater uniformity in performance. Thus was demonstrated the principle of continuous improvement, as shown in the comparison of Figures 3-2 and 3-3.

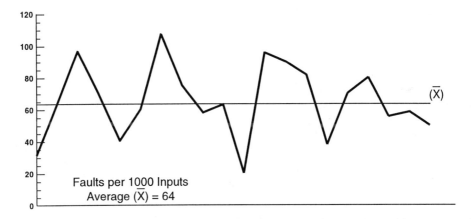

Figure 3-2. Run chart with average number of faults for entire group of trials plotted as straight line.

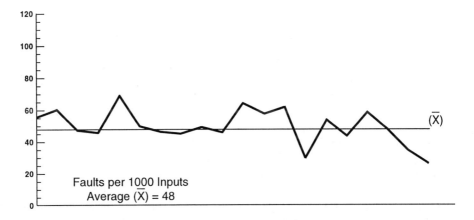

Figure 3-3. Run chart which indicates greater uniformity of performance in test Series No. 100 compared to Test Series No. 1 (Figure 3-2).

The Causes of Deviation

Shewhart's attention was next directed to the dispersion of the data. This involved the distance between the data points and the average, or $\bar{x}$ line. Using a mathematical formula, the standard (average) deviation for the data was determined. Since the range of three standard deviations from either side of $\bar{x}$ will include 99.7 % of the values of a normal population, these became Shewhart's upper and lower control limits, limits derived from the system itself.

A system that is operating within those limits is operating normally for the condition it is in. The random variations that occur within the limits are normal; they are due to the features of the system. These he called "common causes".

However, at times points would fall outside of the control limits, or a series of six or more nonrandom points would appear within the control limits. Shewhart found this to be an indication that an abnormal event had occurred. Something had been different to cause the abnormal reading. He found that the causes of such nonnormal readings were fleeting events, usually easy to identify. These he called "assignable causes". Dr. Deming later renamed them "special causes".

Shewhart had developed the now familiar process control chart. The control chart with normal common cause variations is shown in Figure 3-4. In Figure 3-5 a special cause variation has occurred.

The process control chart did more than identify the nature of the variations; it also measured the frequency of variation occurrences. This resulted in the finding that, in systems that were operating fairly smoothly, 85% or more of the variations that occurred were due to common causes which arose out of the characteristics of the system. The overwhelming potential to minimize undesirable variations and maximize desirable variations therefore lies in improving the system through which the work is done. Changes in the characteristics of the system to effect such changes are clearly the responsibility of management.

The ability to distinguish between common causes and special causes made clear the nature of the work involved in controlling variations in an operating system. Controlling special causes would be easiest. They were usually fleeting events and easily identified, such as the malfunction of a particular machine, a new untrained worker, or a batch of off-grade materials. Much of the time the workers would already be aware of what had happened, and often they would know what needed to be done to correct the situation. Sometimes they would even able to initiate the correction themselves.

The big problem would be controlling common causes, those causes of variation that were built into the system or process. Here, things would change only in response to action to change the system. The workers could do nothing to change such things as poor layout, uneven scheduling, poor maintenance, inferior raw materials or purchased parts, high worker turnover, inadequate training, awkward maneuvering arrangements, unattainable goals, etc. Only through action by management could these causes of undesirable and unwanted variations in the output of the process or system be minimized or eliminated.

An operating system that shows no special causes, no points outside the control limits, and no nonrandom series within the control limits is judged to be operating under statistical control. Adjusting the system should not be attempted, because it will only show up as a special cause variation and make matters worse. Improvement will occur only in response to system change. Dr. Deming often remarked that Shewhart's genius was revealed in his discovery of when to leave a system untouched—when adjustments will only introduce trouble. In retrospect, the method seems

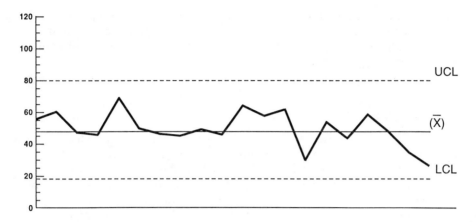

Figure 3-4. Process control chart—variation normal.

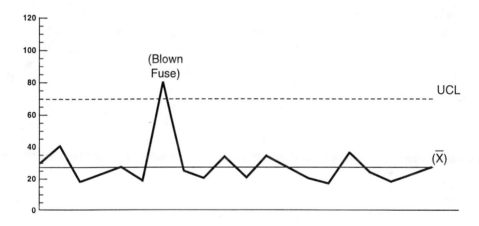

Figure 3-5. Process control chart—special cause variation.

deceptively simple, but it was totally revolutionary. It became the basic building block of the system of managing for continuous improvement.

The Cardinal Principles of Variation

The sophisticated offspring of the run chart, the process control chart, revealed five significant facts:

1. Variation is normal in every system.
2. The causes of variation lie either within the system (common causes), or outside the system (special causes).
3. When a system that is running consistently within its upper and lower control limits, under statistical control, is left untouched, the variations that occur are due to common causes.

4. Common causes arise out of the characteristics of the system, which are determined by management and can only be corrected by management action to improve the system. Workers have no control over common causes.
5. The 85-15 Rule of System Variation: In a normal system, 85% or more of the variations are due to common causes; 15% or less are due to special causes.

Theories Compared

The statistical analysis of the causes of variations in operating systems may now be compared to the observational conclusion of Heinrich's axiom that 88% of work injury accidents are due to unsafe acts, things that people have done wrong.

Heinrich and Shewhart were working at the same time, and there is evidence that Heinrich was aware of Shewhart's work. How then is it that two such opposite theories of variation developed? The answer is not difficult. Shewhart was working for a scientific organization. Heinrich was working for a financial organization. Their charters were totally different.

Shewhart was working to develop a method others could use to achieve the highest levels of reliable service to their companys' customers. He was working on a method that others could use to improve the systems that would provide that service.

Shewhart was working to control the processes and systems that produced the results. His problems were not those of the day, they were those of the future, the future of telephone service and of his company. Shewhart's work was strategic in nature, the improvement of operating systems. His work would affect the way the company was run.

Heinrich's charter was not to research and establish a scientific theory of accident causation. While it is evident in his writings that he was aware of the role of management in accident prevention, his charter was to develop a rationale of accident causation that the insurance carriers could offer their customers without rocking the boat of the customers' prevailing wisdom. (What prospective insurance buyers would give the time of day to a carrier that would tell them they could control accidents and work injuries only by improving their management and production systems, when they knew full well that the root of the problem was the workers?)

Heinrich was working to control the results, the injuries that were occurring. And it was his objective to do it within the framework of the wisdom of the time. Heinrich's work was tactical in nature, affecting only the way things were done. It would not affect the way the company was run.

And he was successful. His philosophy and principles became the foundation of "professional safety engineering." Significant improvement in work injury rates were achieved. Sixty years after their first publication, his ideas remain influential.

But the times are changing everywhere. The new paradigm of the right way to manage is reducing the old thinking to folklore, just as occurred when Newton's new paradigm of the laws of nature appeared. The new way to manage, total quality management, renders the need for a new paradigm of safety imperative. New answers are needed to the question "What is safety all about?" In pursuit of that answer, the value of the required use of statistical methods will become evident.

4

What Is an Accident?
A free-standing event? A variation in a system?
A process within a system, or ... ?

The Need for Definition

The contradictions between the two paradigms of variations in performance is both stark and fundamental, but not irreconcilable. However, reconciling the two paradigms requires constructing a new paradigm of accident causation. It is necessary to recognize that the old paradigm and its corollaries are little more than folklore, built on unsubstantiated axioms and vaguely defined terms.

If you think this characterization of the traditional safety paradigm is harsh, consider the definitions that are offered for our main concern, the accident.

In a review of seven prominent safety textbooks, three did not list *accident* in the index. Of the remaining four, two used the expression "an unwanted incident that causes injury or damage." One defined it as "something that happens that causes...". The National Safety Council's *Accident Prevention Manual for Industrial Operations* defines accidents as "occurrences that may lead to injury, property damage, or both."[1] *The Dictionary of Terms Used in the Safety Profession* provides the following definition:

> *Accident:* An unplanned and sometimes injurious or damaging event that interrupts the normal progress of an activity and is invariably preceded by an unsafe act or condition, or some combination thereof.[2]

William C. Pope, in his book *Management for Performance Perfection*[3] has provided several other definitions contributed by contemporary writers on safety:

> Occurrences of unexpected physical damage to living and nonliving structures. (Haddon, et al.)[4]

> An undesired event that results in physical harm to a person or damage to property. It is usually the direct result of contact with a source of energy...above the threshold limit of the body or structure. (Bird and Loftus).[5]

---

1. Frank E. McElroy, ed., *Accident Prevention Manual for Industrial) Operations*, 8th ed. (Chicago: National Safety Council, 1981).
2. Stanley A. Abercrombie, ed., *The Dictionary of Terms Used in the Safety Profession*, 3rd ed. (Des Plaines, IL: American Society of Safety Engineers, 1988).
3. William C. Pope, *Managing for Performance Perfection, The Changing Emphasis*. (Weaverville, NC: Bonnie Brae Publications, 1990): 107.
4. William Haddon, Jr., Edward A. Suchman, and David Klein, *Accident Research: Methods and Approaches* (New York: Harper, 1964).
5. Frank E. Bird, Jr. and Robert G. Loftus, *Loss Control Management* (Loganville, GA: Loganville Institute, 1976).

> A non-deliberate, unplanned event which may produce undesirable effects, and is preceded by unsafe, unavoidable act(s) and or condition(s). (Thygerson)[6]
>
> The end product of a sequence of acts or events that result in some consequence that is judged to be "undesirable," such as an injury, property damage, interruption, production delay or undue wear and tear. (DeReamer).[7]

It is true that everyone has an understanding of what an accident is, and the textbook examples concur with the dictionary definition. But safety professionals are supposed to be accident prevention specialists, the abolishers of risk, the establishers of safe. Just what is it we are dealing with? What are these unplanned, unwanted occurrences, events or incidents?

A Variation in a System or a Process Within a System?

In Chapter One it was noted that the mandate of Total Quality Management is for every employee to urgently and enthusiastically want to continuously improve the quality and productivity of his or her work: the workers working within the system, the managers and staff working on improving the system with the workers' help; and upper management providing the wherewithal to make the improvements, through concept and policy, budgets and training, staff for assistance, and concern, personal vigor and example to create the conviction and belief in all the people that it can be done.

In order to work on this challenge it is necessary to understand what the work is, what the "products" are, and who the customers are. Clear definitions are essential to this understanding. Everyone—the workers, the supervisors, the department managers, the plant manager, the staff specialists and the executive managers—must know they are talking about the same thing, and all must have the same understanding of what that thing is. And, if possible, it should be a working definition, a guide to action. In order to successfully apply the principles of quality management to safety it is essential to develop a clearer definition of the phenomenon "accident" than presented above. What is an accident? In developing the answer a statistical method will be used.

What are Statistical Methods?

Practitioners of Total Quality Management have strong preferences in the methods they use for finding, analyzing and solving problems. Using statistical methods is viewed as the only way the work can be done, the only way to discover the facts and to develop the understanding necessary to bring about consensus on what the problems are and what should be done.

The use of statistical data, that is facts not contaminated by wishes, hunches, intuition, prejudices, experience or secret agendas, is the foundation that supports the structure of Total Quality Management. It is statistical data, properly selected and interpreted, that identifies the existence of special causes. Finding the special causes

---

6. Anton L. Thygerson, *Concepts and Instruction*, 2nd ed. (Englewood Cliffs, NJ: Prentice, 1976).
7. Russell DeReamer, *Modern Safety and Health Technology* (New York: Wiley, 1980).

and eliminating them leaves the common causes to be identified and worked on. Common causes are more difficult to identify, and it is imperative to approach them on the basis of facts. Quality management by statistical methods is management by facts.

However, not all management systems produce numerical data suitable for mathematical statistical analysis. But the use of statistical methods to find, analyze and solve problems extends beyond mathematical techniques. Some of the most useful tools are simply ways of organizing and displaying data, and others make no use of numerical data at all. They are diagrams that serve to organize thinking. Both the mathematical and the nonmathematical charts and diagrams are included in the concept of statistical methods.

One important nonmathematical statistical tool is the flow chart. A flow chart creates a picture of what happens, or what did happen, or what should happen. It is a diagram of a process or a system, an orderly working totality. In the quest for a definition of "accident," the flow chart is the statistical tool that will prove to be the most helpful. Fortunately, the flow chart that is needed to assist in defining the word "accident" already exists in the literature of safety management. It is the Management Oversight Risk Tree (MORT).[8]

MORT—The Management Oversight Risk Tree

The principal author of the Management Oversight Risk Tree is William G. Johnson, and the principal text is his book, *MORT Safety Assurance Systems,* published in 1980. The first published work on the MORT and associated safety systems was prepared by Johnson as a contract report to the United States Atomic Energy Commission (AEC) in the early 1970s. His 1980 text was written for organizations like the AEC desiring an order-of-magnitude improvement in accident rates and probabilities. A major MORT premise is that the MORT safety system is harmonious with goal-oriented, high-performance, complex management systems. This premise is fully compatible with the mandates of TQM.

The management oversight risk tree might be looked at as an orderly index or catalog of all the factors that constitute the organization's safety culture. However, it might more properly be viewed as a road map, a map that starts with an unwanted destination (the results of an accident) and proceeds from there to examine the possible twists and turns in the safety management system that caused things to go astray.

The management oversight risk tree has two principal working tools. The first is the analytical logic diagram from which MORT derives its name. It is a flow chart, a statistical method using fault tree analysis methodology that traces events and the factors that influenced those events in a tree-and-branch pattern.

The second principal document is the *MORT User's Manual,* which was prepared to help the new user of the analytical technique develop quick familiarity with the MORT process. Together, the logic diagram and the *User's Manual* are the working documents of the MORT analytical method.

---

8. The essential references for the discussions of MORT are: *MORT Users Manual* (DOE 76-45/4, and Safety System Development Center SSDC-4, Revision 2, [1983]) and the *MORT Analytical Logic Diagram,* both published by EG & G Services Inc., 1932 Highway 92W, Suite D, Woodstock, GA 30188; and William G. Johnson's book, *MORT Safety Assurance Systems* (New York: Marcel Dekker Inc., 1980).

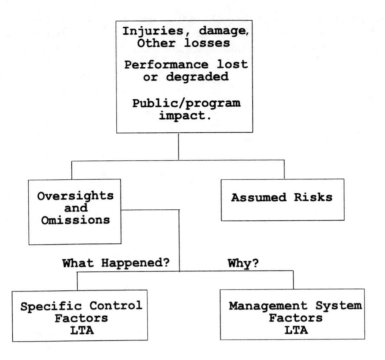

Figure 4-1. The Management Oversight Risk Tree (MORT).

The MORT Top Events

The basic theory of the management oversight risk tree is set forth in the top events of the MORT analytical logic diagram, as shown in Figure 4-1. The initiating question is "What and how large were the losses?"

The management oversight risk tree uses the past tense, because it is a look back at what happened. It works from the result (the losses) back to the causes. The basic theory upon which the MORT is constructed is simple. Risk losses are caused either by:

1. Management planning or operational oversights and omissions, or by
2. Risks that, after due consideration at the appropriate level, management had consciously assumed.

These postulates are congruent with Shewhart's findings on system variations. Those findings were:

1. Variation is normal in every system.
2. The causes of variation lie either within the system, (common causes) or outside the system (special causes).
3. When a system that is under statistical control (performing consistently within its upper and lower control limits) is left untouched, the variations that occur are due to common causes.
4. Common causes arise out of the characteristics of the system which are largely determined by management, and can only be corrected by management action to improve the system.

5. In systems that are operating fairly smoothly, 85% or more of the variations that occur are due to common causes, and 15% or less are due to special causes.

Shewhart's control chart revealed that 85% or more of the variations that occur in an operating system have their origins in the characteristics of the process or system. The first MORT postulate identifies management planning or operational oversights and omissions as the source of those characteristics in the accident sequence. The accident sequence is an ongoing process, albeit an unwanted one.

The second postulate of the MORT theory is also congruent with Shewhart's findings, specifically the finding that 15% or less of the variations that occur are due to special causes, the fleeting events that are introduced into the system in one manner or another. Very few of the loss incidents that occur are the result of a conscious, reasoned management decision at the appropriate level to accept the risk, that is subsequently proven to have been wrong.

In the MORT theory, risk losses that cannot be attributed to a considered assumption of risk are considered to be the result of management planning or operational oversights and omissions. A risk that is unidentified is judged to be a management oversight or omission, just as an identified risk that is not acted upon. Any ensuing mishap will have arisen from the characteristics of the system, a common cause.

An assumed risk that resulted in a mishap would be regarded as due to a special cause, a singular event introduced into the system, as opposed to being a variation within the system. (The decision to fly in the Challenger disaster is an example of an assumed risk that went awry. As such, it cannot be classified as an accident.)

In summary, the control chart revealed that variations in operating systems are normal, and that 85% or more of the variations that occur have their origins in the characteristics of the system. People working within the system can do nothing about these characteristics. The management oversight risk tree identifies management oversights and omissions as the source of these characteristics or common causes in the accident sequence, a sequence that is an ongoing process when it occurs.

This is fundamental change, not only in safety management strategy, but also in organization culture. No longer will workers be continually reminded that they are responsible for 85% or more of the accidents that occur, something they have known all along was not true. Responsible for some accidents, yes, but for 85% of them? No! Consider their relief to find that they will no longer be criticized for things they have no control over: the characteristics of the tasks or the work environment that they can do nothing about. Primary responsibility for incidents can be assessed to the worker only after planning and operational oversights and omissions have been responsibly judged not to have been responsible.

And what about the supervisors? With everyone geared to system improvement, they will no longer be inclined to play the coverup game. In fact, it will be to their detriment to do so. Fear of criticism or reprimand for unwanted incidents will have been removed. Honest reporting and diligent investigation and analysis, with the assistance of everyone involved, will be nurtured.

What about executive management? One of the defining objectives of quality management is the elimination of wastes, the waste of assets, the waste of materials and the waste of the time of people, all people, all levels. The wastes that arise out of accidents are wastes of all of these things. Establishing the elimination of wastes that arise out of accidents as a defining objective for total quality safety management establishes safety's economic role in the organization, a role that executive management in quality-managed organizations will expect and, in fact, demand.

The traditional safety paradigm that largely restricts safety's objectives to the prevention of injuries and damage (controlling the results of accidents) will not adequately serve the economic role for quality safety management, although, as we shall see, controlling the results of accidents will continue to be important.

In the top events of the MORT, the basic question was "What and how large were the losses?" In Total Quality Safety Management, the question should be rephrased to "What and how large were the wastes?"

In the MORT basic theory, incidents that are not the result of a conscious assumption of risk are seen to arise from management planning or operational oversights and omissions. Only they can properly be called "accidents." However, this understanding does not provide a working definition of an accident.

What Happened and Why?

In order to determine whether an accident was due to a considered and calculated assumption of risk, or due to management planning or operational oversights and omissions, the questions become, "What happened?" and "Why?" Each question becomes a major branch of the MORT tree.

"What happened" is attributed to the inadequacy of specific control factors. "Why" these specific control factors were less than adequate is attributed to management system factors that were less than adequate. This relationship is presented as the last two factors of the top events of the management oversight risk tree. Obviously, the accident itself rests in the "What Happened" branch of the oversight risk tree. At this point we are up against the "something that happens," the "unwanted incident or event," "the occurrence that causes..." of the definitions cited earlier. But what is the "something," the "incident," the "event," or the "occurrence"?

The Accident Triad

Strangely enough, the answer was present in the original dominoes, only it was either missed or disregarded. In presentation the dominoes seem to fall by themselves. But, do they really? Of course not. The demonstrator has to give Domino No. 1 a push. Energy is needed, and it is that flow or transfer of energy that creates the "something that happened," the "unwanted incident," or the "occurrence that causes."

Unwanted energy transfer is involved in every accident. In fact, it is the identifying element in the word "INCIDENT." For the purposes of this discussion, however, it is necessary to employ the broad, all-inclusive understanding of that word, extending it beyond that of a free-standing event or occurrence. It must include the development of, or exposure to, a hostile environment such as excessive noise, extreme heat or cold, noxious or oxygen deficient atmospheres, etc. It must include the existence of work conditions that exceed the tolerance level of the human body or psyche, such as heavy lifting, harmful repetitive motions, conditions that induce physical deterioration; unreasonable output expectations, or supervision by fear, conditions that result in excessive mental stress for average, healthy workers. These chronic exposure potentials must be included with the exposure to acute events in the understanding of the concept of unwanted energy transfer or "incident."

These difficulties with the word "incident" have been recognized by Fred Manuele in the discussion of causation models in his book *On The Practice of Safety*.[9] Manuele neatly solves the dilemma with the suggestion for a new word, a word that circumscribes these related elements into a single concept. His suggested word is **HAZRIN**—a compounding of HAZard Related INcidents.

Manuele writes: "HAZRIN as a term encompasses all incidents that are the realization of the potential for harm or damage, whether harm or damage resulted or could have resulted for all fields of endeavor that are hazard related." He notes that HAZRIN is a word that emphasizes *possible consequence*. The value of this contribution is recognized here by replacing the general term "incident" used in the MORT with the new word.

However, in adopting the new term, Manuele's phrase "whether harm or damage resulted or could have resulted" has been omitted. The understanding employed here is that: "HAZRIN is a term that encompasses all incidents that are the *realization* of the potential for harm." The purpose of this deletion is to hold the definition to and emphasize *possible* consequences.

But this restriction does more than simply emphasize possible consequence. Without this restriction a hazard related incident, a HAZRIN, is not an accident—there have been no losses. The worst that has happened has been a close call or a "near-hit". It was a close call simply because no persons or objects were in the path of the energy flow or exposed to the environmental condition. In the domino sequence this can be illustrated by simply removing Domino No. 5, or moving it to one side. Removing the domino demonstrates one important type of safety control—evacuation; moving it to one side is evasion.

There is another way to protect Domino No. 5. Simply place a barrier between dominoes No. 4 and No. 5. But, if that barrier is not adequate, and a control procedure such as evasion or evacuation is not possible, trouble is ahead. A far better answer would be to control the potential for harm, the unwanted energy transfer, or hostile environmental condition initially.

However, when:

1. a potentially harmful energy transfer or other hazard related incident occurs, *and*
2. the barriers or controls are less than adequate, *and*
3. persons or objects are in the path of the energy flow, or exposed to the environmental condition,

the **"ACCIDENT TRIAD"** is complete. The accident will occur and wastes (injury, damage or degraded performance) will result. This relationship is shown in Figure 4-2.

Operational Definition of "Accident"

At this point the need is for an operational definition of *accident*. "An operational definition puts communicable meaning into a concept... . It is one that reasonable men can agree on ... one that people can do business with ... expressed in operational terms of sampling, test and criterion."[10]

The need is for an operational definition that recognizes the three elements of the Accident Triad and their origin in oversights and omissions.

> An *accident* is a process within a system. It is an unwanted transfer or flow of energy that, due to barriers and/or controls that are less than adequate,

---

9. Fred A. Manuele, *On the Practice of Safety* (New York: Van Nostrand Reinhold, 1993).
10. Deming, *Out of the Crisis* (MIT-CAES): 276-77

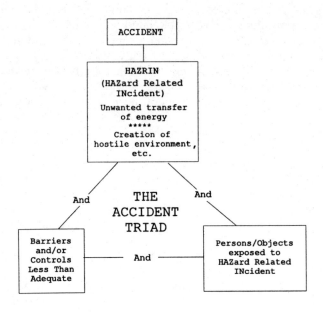

Figure 4-2. The Accident Triad.

result in harm to the persons or objects in the path or exposed to the unwanted transfer.

What Safety Is All About

Safety is all about the management of exposure to the myriad expressions of energy that confront all continuously. The aim is to derive benefit from these expressions and avoid, or at least minimize, harm. The concept is the same whether the energy involved is the pull of gravity on the icy sidewalk, the intimate contact with the energy of the blood-borne pathogen, or the deprivation of the energy oxygen so essential to life.

In the workplace the focus of managing safety is preplanning the management of energy for high performance without damaging, unwanted energy transfers or other hazard related incidents, including the creation of hostile environmental conditions.

5

The Tactics and Strategies of Safety Management

In the last chapter it was noted that the management oversight risk tree might be looked at as an index or catalog of all the factors that constitute an organization's safety culture, the total pattern of behavior as it relates to the matter of safety and accidents. The MORT was conceived as a technique for thorough, searching investigations of occupational accidents and analysis of safety programs.[1] In creating MORT, Dr. William G. Johnson worked with teams of safety professionals at advanced technology facilities where workers were exposed to very high energy levels. As a result, MORT has an appearance that many consider unnecessarily complex. It literally seems that the author(s) left no stone that could be found unturned in the search.

In many ways, this complexity has worked to the disadvantage of more universal acceptance of the theories and disciplines involved. "They've got to simplify it." is an expression often heard in discussions among safety professionals concerned with the more mundane safety problems of ordinary exposures. While this reaction to the MORT is understandable, it overlooks the fact that it is not the MORT that is complex, but rather the accident phenomena.

Rather than an *index*, a better analogy for the MORT is a *roadmap*. A map of all the roads in the United States is also a very complex diagram, but it serves well in getting the traveler from point A to point B, even though there are innumerable ways to make the trip. You only have to select the proper roads, not try to follow them all. So it is with the MORT. The proper roads to the definition of an accident were found without traveling the entire map.

Determining the Destination

The first requirement of any trip is to determine where you want to go. So it is with the MORT. We have already seen one major fork in the road, with the path to the left leading to the "Land of What Happened" and the path on the right leading to "why?". In the vernacular of the MORT, the "what happened" route is identified as the *specific control factors* path, and the "why" route is identified as the *management system factors* path. Early in the journey there is no choice involved. The "Land of What Happened" must be visited before the visit to the "Land of Why" will have any meaning.

---

1. William G. Johnson, *MORT Safety Assurance Systems* (New York: Marcel Dekker, Inc., 1980): v

The Anatomy of an Accident

We have been traveling the "what happened" route up to this point. On this trip we have found that an accident, when it occurs, is made up of three concurrent phenomena: (1) there is an unwanted transfer of energy or environmental condition, a *hazard related incident*; (2) barriers and/or controls are less than adequate, and (3) there are persons or objects exposed to the HAZRIN. This establishes the *accident triad*, which enables an accident to be viewed as a process that contrasts sharply with the process of falling dominoes. This new view also identifies "safety" as the management of unwanted energy transfers or environmental conditions through the use of barriers and/or controls. This view can now be expanded, as shown in the flow chart in Figure 5-1, The Anatomy of an Accident.

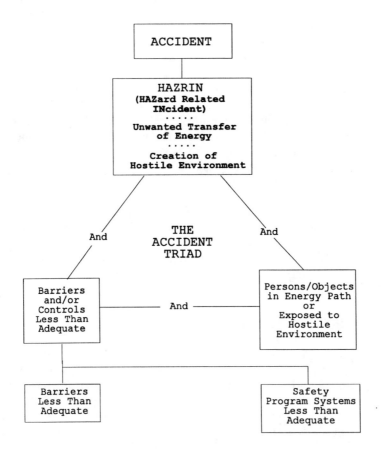

Figure 5-1. The anatomy of an accident.

In Figure 5-1, the portion of the accident triad labeled "Barriers and/or Controls Less Than Adequate (LTA)" is separated into two activity channels. "Controls Less Than Adequate" has been retitled "Safety Program Systems Less Than Adequate" to more accurately describe the nature of the controls. This separation indicates there are two different approaches available for use in preplanning the management of hazard related incidents. From the recognition that two different management approaches exist flow a whole series of options as shown in Table 5-1. The distinction is fundamental. It is the second fork in the road.

The use of barriers to render conditions safe also constitutes a safety program system, but it is a system totally different in purpose and character from the program elements entitled Safety Program Systems. The Barrier Systems are concerned with *controlling the results of the incident—protecting the targets from harm.* Safety Program Systems are concerned with *preventing the incident—improving the systems that produced the results.*

These differences affect the nature of the work involved in achieving continuous improvement in the quality and productivity of our work. This is made clear in a comparison of the work involved in managing barriers with that of managing safety program systems.

In the pursuit of continuous improvement in the quality and productivity of safety systems, these distinctions between tactics and strategies are critical. Continuous improvement in both areas is important, but the greater promise lies in the improvement of safety program systems—the strategic programs. However, improvement in the management of barrier systems—the prevention of injuries or damage—will always be one of the primary concerns of a safety program. That will be the first route to explore on this portion of the trip.

Table 5-1. Barriers vs. Safety Program Systems—Tactics vs. Strategies

Barriers	*Safety Program Systems*
The problems are tactical. They affect the way the task is done and under what conditions.	The problems are strategic. They affect the nature of the operation.
Departures from safety are the concern.	Root causes are the concern.
Diagnosis is needed to identify the problem.	Analysis is needed to identify the problem.
Solving the problem is the task.	Defining the task is the problem.
Providing an answer is the need.	Improving the system is the need.
Corrective action will be a project.	Corrective action will be a process.
Change will be physical or procedural.	Change will be mental, a new way of thinking, a new strategy.
Cost may affect alternatives.	Economics is not a problem.
The solution must meet a specific requirement—make the task safe.	The solution must achieve the goal of system improvement.
Developing technical alternatives needs input from all.	Developing operational alternatives needs management leadership.
Implementation will be by consensus decision.	Implementation will require management action for change.

6

The Concepts of Energy and Barriers
Controlling the capacity to do work or harm

The Theory of Controlling Results

Concern about workers' injuries and deaths due to accidents or exposure to environmental hazards is found in ancient writings. Measures taken to protect workers are also mentioned. The Roman naturalist Pliny the Elder, in the first century AD, wrote of using ox bladders as respirators to protect workers against mercury fumes in the production of vermilion. He also wrote of the sickness of the lungs of slaves whose task was to weave asbestos into cloth.[1]

From those early days to recent times, concerns about casualties in the workplace have centered on controlling the injuries and deaths resulting from accidents. Prior to 1930 such efforts were largely unstructured, without the benefit of an organized theory.

It is probable that Heinrich's greatest contribution to safety professionalism was providing the first widely-accepted theory of accident causation. This is reflected in references to him as the "father of modern safety engineering". After the publication of his theory, the body of safety information developed into a library of safety knowledge of tremendous proportions. His contribution was not the cause of the information explosion, but it aided rational, organized thought. However, the primary concern—preventing injuries, controlling the results of accidents—remained unchanged.

In the 1960s, changes in causation theory were occurring. In 1967 Ross McFarland made the following comments in a paper presented at the National Safety Congress:[2]

"All accidental injuries result (1) from the applications of specific forms of energy in amounts exceeding the resistance of the tissue upon which they impinge, or when there is interference in the normal exchange of energy between the organism and the environment (e.g., as in suffocation by drowning). Thus the various forms of energy...constitute the direct causes of injuries in accidents. Also, prevention of injuries can often be achieved by controlling the source of energy, or the vehicles or carriers through which energy reaches the body.

While the specific types of energy which give rise to injuries are quite limited in number, the forms in which they abound and the variety of the vehicles or carriers of energy are innumerable. Man himself is constantly compounding

---

1. Paul Brodeur, "Annals of Law, The Asbestos Industry on Trial" (*The New Yorker*, June 10, 1985).
2. Ross A. McFarland, "Application of Human Factors Engineering to Safety Engineering Problems" (Paper presented at the National Safety Council Congress, *National Safety Council Transactions*, 1967).

this situation as he develops more powerful sources of energy and puts various kinds of energy to new uses."

While McFarland's eye was still fixed on the results of the energy flow, he pointed out, almost in passing, that the "...prevention of injuries can often be achieved by controlling the source of the energy, or the vehicles or carriers through which the energy reaches the body." It should be noted that McFarland's theory applies to damage events (or incidents) as well as bodily injury.

This was a fundamental change in accident causation theory. While the concept was not original with McFarland, his statement was valuable in directing attention to the role of energy in accidents. A new tactical target for safety was identified, and a new theory of how to make tasks safe: "...controlling the source of the energy, or the vehicles or carriers through which the energy reaches the body (or structure)." The need now was for increased understanding of energy and the techniques for controlling it.

The techniques of controlling energy were already well developed. However, at this point, we need to direct our attention to the understanding of energy.

Energy Types and Sources

Energy is defined as the physical capacity to do work. It is essential to performance. The abnormal, unexpected, or unwanted transfer or flow of energy causes accidents, and results, not in productive work, but in wasteful harm. As McFarland points out, the different types of energy that cause accidents are relatively limited, but their sources are practically unlimited. Some types are encountered only rarely, while other types are all-invasive and constantly present. The following discussion is not intended to be scientific or encyclopedic. The purpose is to explore the low ratio of types of energy to the almost unlimited sources of energy that can cause accidents.

Gravity

The most common energy source is *gravity*. It is elemental and all-pervasive, except in space. It expresses itself in the form of mass or weight, and involves the dimension of height. It is gravity that energizes falls and falling objects, building collapses, airplanes crashes, the water pressure in elevated tanks, and the speed of roller coasters.

Kinetic Energy: Motion

The second most common type of energy is *kinetic energy* or *motion*. Kinetic energy expresses itself in two different forms, *inline* motion or *rotating* motion. Kinetic energy also has two primary sources, *human* and *mechanical*. Human linear kinetic energy is found in such actions as walking, running, jumping, climbing up or down, pushing, pulling, lifting, carrying, reaching, stretching, etc., and in such reciprocating actions as using a hand saw, pumping a bicycle, or rowing a boat.

Human rotational motions most commonly involve the bending or twisting of the body or its parts, such as the wrists, neck, or back. As a result of the flexibility of the human body, combinations of inline and rotational motions are common.

Bodily injury from kinetic energy can be either acute or chronic in nature. Historically, safety has been primarily concerned with the sudden or acute injuries resulting from release of kinetic energy. In recent years, however, this limited concept of trauma has been expanded dramatically through advances in medical science. These advances have led to greater understanding of the nature of trauma that is chronic, such as injuries that arise from repetitive motions. These motions are now recognized as exposures to low-grade levels of energy transfers that result in cumulative trauma. The medical advances have in turn resulted in judicial recognition of cumulative trauma as compensable. Cumulative trauma injury is generally referred to as an illness.

Mechanical inline motions include moving vehicles of every type: cars, trucks, planes, trains, boats, etc. It also finds expression in mechanical reciprocating actions: the up-and-down, back-and-forth, in-and-out, and clamping action of machines. Examples include pistons and cylinders, elevators, hoists and cranes, vibrators and shakers, punching and shearing machines, pushers and stackers, etc. Forklift trucks combine inline and reciprocating actions.

Mechanical rotational motions are common indeed, and are found in wheels, gears, pulleys, shafts, circular saws, drills, lathes, slicers, grinders, augers and screw conveyors, fans and blowers, mixers and blenders, rollers, centrifuges, etc. Many machines are a combination of linear reciprocal and rotational action: surface planers or dough breaks, for example. Moving vehicles combine the rotational action of the wheels and the linear action of the vehicle itself.

Combinations of energy flows in one machine or system are common. The forklift truck not only involves reciprocating and inline motion as mentioned, it also has the rotational action of its wheels. Robotics can be designed to combine a multiplicity of motions.

Other Types of Energy

Gravity and motion are the two most common types of energy. However, many other forms of energy—other capacities to do work or harm—exist. Consider the following:

- **Pressure.** Air, hydraulic, steam, gases, etc., found in lines, tanks, compressors, pumps, boilers, autoclaves, and other closed systems. Noise is acoustical pressure in an open system.
- **Stresses.** Tension, vibration (harmonics), coiled springs, suspended loads, etc.
- **Electricity.** High voltage, low voltage, static. Present in transformers, generators, battery banks, service outlets and fittings, motors, heaters, power tools, lightning, etc.
- **Chemicals.** *Explosives*—blasting agents, flammable liquids, oxidizing-reducing agent combinations, dusts, hydrogen, propane, and other gases. *Corrosives*—acids and caustics. *Toxins*—chlorine and chlorine compounds, carbon monoxide, lead and lead compounds, asbestos, silica, pesticides and herbicides, "natural chemicals" such as the toxins in insect and poisonous snake bites, allergens such as poison ivy, poison oak, etc.
- **Radiation.** Ionizing (nuclear) and non-ionizing, solar, X-rays, lasers, radar, ultraviolet, infrared, etc. Arc welding, magniflux, etc.
- **Thermal.** *High temperatures*—fire, ovens, furnaces, torches, crucibles, exposed steam pipes, atmospheric, etc. *Low temperatures*—cryogenics, liquefied gases, blast freezers and freezer storages, atmospheric, etc.

Interference with Normal Energy Transfer

The preceding discussion is an overview of energy transfers that have the potential to do harm when unwanted transfers occur. At this point it is necessary to consider the hazard potential that lies in the interruption or failure of energy transfers to occur, transfers that are critical to life.

All living things are dependent on beneficial energy transfers of some type. When such transfers are interrupted or fail, the life process is at great risk. For animate objects, man included, the most vital transfer is for an adequate and continuous supply of oxygen.

The understanding of the phrase "unwanted transfer of energy or creation of a hostile environment" must include the interruption or failure of an adequate supply of the chemical energy that oxygen provides all animate forms of life. Death by asphyxiation will result. Death by dehydration or starvation are related phenomena.

Extending the Understanding of "Creation of a Hostile Environment"

The foregoing discussion of unwanted transfers of energy is primarily based upon hazard related incidents that result in discrete traumatic incidents. However, the inclusion of the phrase "creation of a hostile environment" in the definition has broad implications that are rarely commented upon, although the technical specialty of "safety" is often referred to today as "safety and health."

The recorded history of mankind is replete with "incidents" of plagues, which occur continually on one scale or another throughout the world even to this day. Does the outbreak of a disease epidemic result from the existence of a "hostile environment"? Does the transfer of disease-causing organisms constitute "an unwanted transfer of energy"? Is the logic of incident causation by energy transfer or by creation of a hostile environment the same? Are the measures of controlling unwanted energy transfers relevant to the prevention and control of the health problems that arise from hostile environments? For this author the answer is "yes." The principal difference would appear to be one of pace or time, the acute or sudden, the chronic or slowly evolving. This opinion is supported by the marriage between "safety" and "health." Are the control measures for each, the acute and the chronic, related? These control measures are the next consideration.

It is not unusual for two or more kinds of energy to be involved in a hazard-related incident. Johnson has noted that the careful and precise tracing of energy flows in the investigation of an accident, or in the planning and design of a new process, or in the conduct of a safety audit or inspection, has proven to be a valuable approach in accident analyses and accident potential recognition. He points out that hazardous combinations of energies are more likely to be recognized, and that additional opportunities to interrupt sequences are more easily identified.

The Tactics of Results Management by Barriers

Johnson credits James J. Gibson[3] and William Haddon, Jr.[4] with the development of the concepts of energy and barriers. In the *MORT User's Manual* glossary, barriers are defined as "physical or procedural measures to direct and maintain energy in wanted channels and control unwanted release."

In the Management Oversight Risk Tree, four systems for the use of barriers are presented:

---

3. James J. Gibson, *Behavioral Approach to Accident Research* (Chicago: National Society for Crippled Children and Adults, 1961).
4. William Haddon, Jr., *The Prevention of Accidents, Preventive Medicine* (Boston: Little, Brown, 1966).

1. Barriers on the source of the energy flow or the hostile environment.
2. Barriers between the source and the targets—the persons or objects in the path of the energy flow, or exposed to the hostile environment.
3. Barriers on the targets—the persons or objects so exposed.
4. Separating the targets and the energy flow or hostile environment by time or space.

Haddon's concepts of energy and barriers included tactics for the management of unwanted energy transfers and, where applicable, they were presented in order of preference for implementation. Figure 6-1 is an adapted version of these strategies as Johnson presents them.[5]

Figure 6-2 is adapted from a flow diagram developed by Johnson which depicts the process sequences of the control of energy.[6]

There is nothing new or unfamiliar in the tactical measures of control listed as examples. Some measures are physical in nature and others are procedural. All are standard practice in improving workplace safety through the control of harmful contact with unwanted transfers of energy, by improving how things are done and under what conditions. Others might be added.

What is new is arranging the tactical measures into systems of energy control strategies and placing them in a suggested progressive order which can be used in various combinations. Johnson has pointed out, however, that in application the rank order may change. It is easier, for example, to protect drivers from injury through the use of seatbelts and airbags, than it is to limit or prevent the buildup of energy in the moving vehicle.

Johnson has noted, however, that as the level of energy escalates, the need for redundant, successive strategies and barriers increases. The "principle," that escalating energy levels should be matched with an escalation of safety efforts, has long been honored in practice but not discussed. The general observation that the best and most effective safety programs are found in operations involving high energy processes is testimony to that relationship. Johnson makes the further point that systematic review of available energy control tactics and the creation of optimum mixes to reduce harm has not been customary in safety.

The systems of process sequences for controlling energy constitute safety systems of Results Management by Barriers. These systems are displayed in diagrams which expand the material presented in Figure 6-1 by presenting representative tactics that are familiar to everyone. They are presented in the priority order suggested by Haddon. Figure 6-3 is concerned with actions to control unwanted energy transfers.

The tactics of barriers on the energy source are related to Haddon's comment that "...prevention of injuries can often be achieved by controlling the source of energy...." The examples presented here are only representative, and the distinctions between tactics are arbitrary and imprecise. Combinations are often productive. Imagination in developing successful barrier tactics needs encouragement and broad inputs.

It should be noted that the safety systems that center on the energy source apply not only to acute unwanted energy transfers but to chronic transfers as well. These long-term, low-level transfers result in such disorders as low back pain and upper extremity repetitive trauma. The results of such "acute transfers" on the jobsite are referred to as *occupational injuries*, while the results of chronic transfers are referred to as *occupational illnesses*. The practices of control are referred to as ergonomics. Injuries

---

5. Johnson, *MORT Safety Assurance Systems* (New York: Marcel Dekker, Inc., 1980): 29.
6. Ibid.: 32.

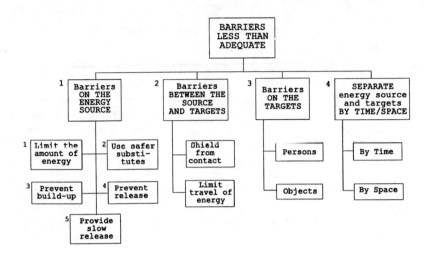

Figure 6-1. The tactics of results management by barriers.

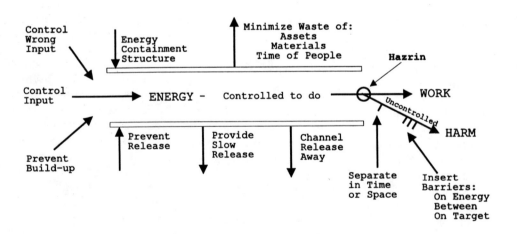

Figure 6-2. Process sequences of controlling energy—the ability to do work or harm.

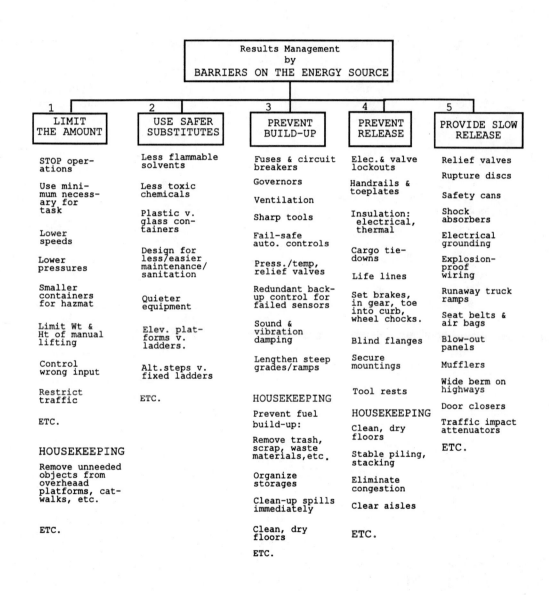

Figure 6-3. Actions to control unwanted energy transfers.

arise from energy sources that are external, while the illnesses more often arise from transfers that occur internally, the result of excessive strain, repeated unnatural twisting or stretching, prolonged unnatural body positioning, or other harmful applications of human energy. The control principles are the same for both.

The Culture of "Good Housekeeping"

It also should be noted that the generalized tactic of *housekeeping* appears under three different approaches to barriers on the energy source: (1) limiting the amount of energy; (2) preventing the buildup of energy; and (3) preventing the unwanted release of energy. The role of housekeeping in the safety program is rarely discussed, and has never been clearly defined. Yet the housekeeping conditions found in any operation are widely considered to be a meaningful barometer of safety-mindedness.

The benign academic neglect the subject of housekeeping has received is both caused by and results from the fact that the concept has no real conceptual home. It is neither an engineering concept nor a management concept. It is a cultural concept, a state of mind. And, it is reflective of the nature and personality of the ongoing endeavor, whatever that endeavor might be. This is precisely why the quality of the housekeeping serves well as a barometer of the safety-mindedness of the operation.

Good housekeeping is a cultural concept that recognizes the importance of the little things that too often sire bigger things. It pays attention to the beginning of things that ought never happen. It nurtures attention to details—details that have the potential to create unwanted problems if left unchecked. Good housekeeping is proactive in the best sense of the word—it is not reactive.

The creation and maintenance of high standards of housekeeping creates a culture in the workplace where "good enough" is not acceptable, where only "the best" and "the right way" will be the cultural creed. Pride in the job, product or service, in the organization, and in your own contribution and self will be nurtured. Good housekeeping is of the essence of employee involvement, and of quality safety management. One comment often made states that "People will deliver to your level of acceptance, not your demand level. If you demand one thing and accept less people will deliver less." Not by coincidence, these remarks on the nature of good housekeeping apply equally to the matters of good sanitation and good personal hygiene.

The three remaining tactics in Haddon's classification are more closely identified with the second phrase in his statement, which refers to action involving "the vehicles or carriers through which energy reaches the body (or object)." They are directed to protection from an unwanted transfer.

Haddon's second-ranked barrier tactic is "Barriers Between the Energy Source and the Targets"—the persons or objects in the path of the unwanted energy flow or exposed to the hostile environmental condition. The concept is diagrammed in Figure 6-4.

The third and fourth barrier concepts, "Barriers on the Targets" and "Separation of the Targets and the Energy Source," are shown in Figures 6-5 and 6-6.

The MORT Analytical Logic of the Accident Triad

The Management Oversight Risk Tree presents two concepts:

1. It presents the total safety program as a specialized management subsystem that is focused on organized, programmed, control of workplace safety hazards;
2. It displays a structured set of interrelated safety program elements and concepts as a universal logic diagram that becomes a master "work sheet" for use in analyzing a specific accident, or for evaluating and appraising the existing safety program for hazard-related incident/accident potential.

As a safety management program MORT has been designed to:

1. Prevent safety-related oversights, discrepancies (errors), and omissions;
2. Enable the identification, assessment, and subsequent referral of residual risks to the proper levels of management for appropriate action;

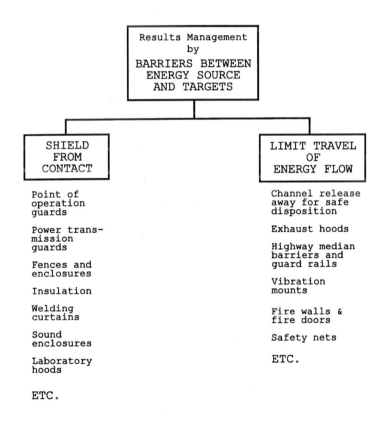

Figure 6-4. Barriers between energy source and targets.

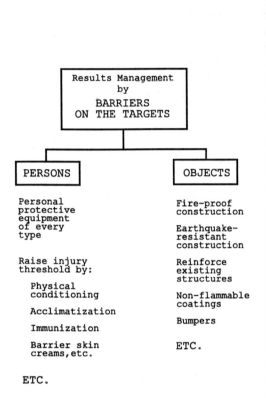

Figure 6-5. Barriers on the target.

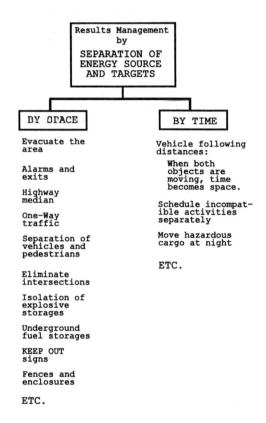

Figure 6-6. Separation of the targets and the energy sources.

3. Establish priorities for the allocation of the resources available to the safety program and to individual hazard control efforts.

The Accident Triad identifies the elements of the accident sequence that are subject to analysis for cause. The logic for analysis of each system in the diagram are presented in Figures 6-7, -8, and -9. The questions excerpted directly from the *MORT User's Manual* appear in italics.

Potentially Harmful Incident (HAZRIN) (Figure 6-7)

What was the Hazard Related Incident, the energy flow or environmental condition that resulted in the accident?

Figure 6-7 denotes an energy flow or environmental condition which *could* result in harm if barriers and safety program systems are inadequate and a vulnerable person or object is exposed.

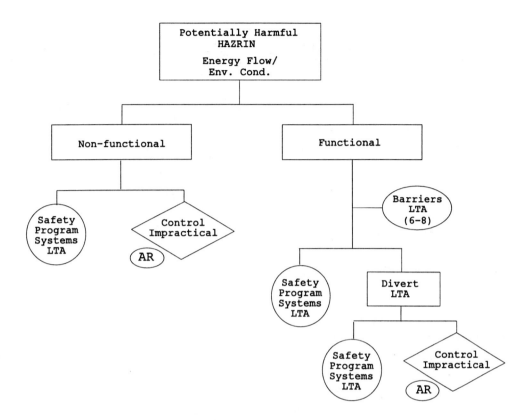

Figure 6-7. Potentially harmful incident (HAZRIN).

NONFUNCTIONAL

Was the energy flow or environmental condition causing the harm to a functional part of the product or system? Was there adequate control of nonfunctional energy flows and environmental conditions? Was such control practicable?

Note that the event is flagged "AR" as an "Assumed Risk." Proper management level must assume responsibility for this decision.

FUNCTIONAL

Consider the lower-tier elements below this only if the HAZRIN was a functional part of or a product of the system. Given a failure of the barrier system:

Were the administrative controls adequate to prevent the harmful energy flow or environmental condition from reaching vulnerable persons or objects?

Was diversion impractical?

Note that the event is flagged "AR" as an Assumed Risk. The proper management level must assume responsibility for this decision.

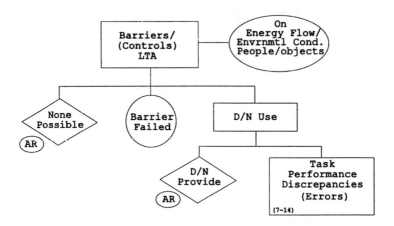

Figure 6-8. Barriers and safety program systems (controls) LTA.

Barriers and Safety Program Systems (Controls) LTA (Figure 6-8)

Were adequate barriers and safety program systems in place to prevent vulnerable persons and objects from being exposed to harmful energy flow or environmental conditions?

Note: The constraint placed on 6-8 is intended as a device to prevent oversight. It is designed primarily to draw attention to barriers and control related to harmful energy flows or environmental conditions and those controls designed to control movement of target persons or objects.

Both types of barriers should be considered, but rigorous and proper classification is not necessary to the analytical process, provided that all barriers are considered.

Were the barriers and controls (safety program systems) designed to prevent harmful energy flows or environmental condition from reaching vulnerable people or object less than adequate? (Refer to Figures 6-8 and 6-10 for further development.)

Were barriers and controls (safety program systems) designed to prevent vulnerable people and objects from encountering harmful energy flows or environmental conditions less than adequate? (Refer to Figures 6-8 and 6-10 for further development.)

Vulnerable People or Objects (Figure 6-9)

Note: The constraint "value" in place here. An accident is defined in terms of loss of something of "value."

What vulnerable people and/or objects of value were exposed to the harmful energy flow or environmental condition?

NON-FUNCTIONAL

Was the person or object performing a functional role in operation of the system? Was such control practicable?

Note that the event is flagged with an AR symbol. The proper management level must assume risk responsibility for this decision.

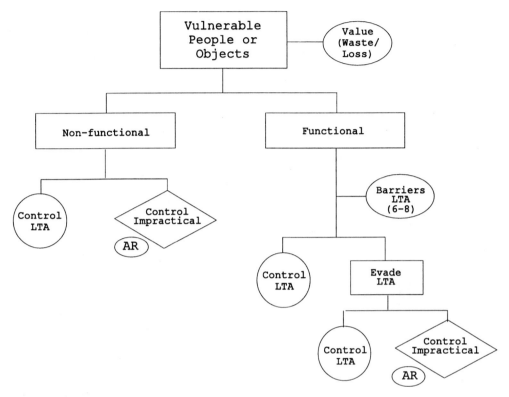

Figure 6-9. Vulnerable people or objects.

FUNCTIONAL

Consider the lower-tier elements below this only if the person or object was performing a functional role in the operation of the system. Given a failure of the barrier system:

Were the administrative controls (safety program systems) adequate to prevent persons or objects from being exposed to the harmful energy flow or environmental condition?

Note: The constraint in place here. An accident can occur only if the barriers were LTA.

EVASIVE ACTION LTA

Was there adequate evasive action for vulnerable persons or objects? Was evasion impractical?

Note that even this is flagged with an AR symbol. An appropriate management level should assume risk responsibility for this decision.

Figure 6-10 extends the MORT diagram to include the Tactics of Results Management.

The Day-To-Day Safety Program Activities

On a day-to-day basis, safety activities on the job are largely devoted to the application and monitoring of energy management by barriers in ongoing operations.

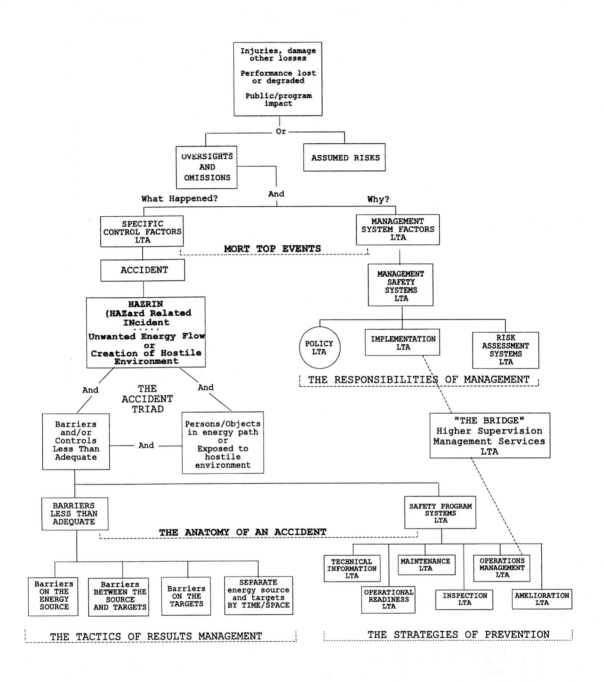

Figure 6-10. The Management Oversight Risk Tree: What and how large were the wastes?

Day-to-day safety activities are closely tied to employee actions—compliance with prescribed job procedures, using protective equipment, keeping safeguards in place, and good housekeeping practices. They are related to ensuring that machines and equipment are in safe working order, that workers are not exposed to falling, being struck, caught, or exposed to contact with electric current or hazardous materials, subject to overexertion, and so on. All of these activities are essentially a response to the prevention of injury, to controlling the results of unwanted energy transfers.

With this in mind, it follows that the better understanding everyone has of the nature of energy, its pervasiveness and its ability to do harm as well as work, and of the concepts of controlling energy by barriers, the more effective will be the control of the results of unwanted transfers in daily living. Everyone needs to be aware of, understand, and respect the energy characteristics of their environment, not only on the job, but off the job as well. This is basic to effective hazard and accident potential recognition.

If the objective is improving the application and monitoring of the systems of barriers to control the results of unwanted energy flows in the workplace, certain questions become pertinent:

> Who is most intimately familiar on a constant basis with incidents of unwanted energy transfers?
>
> Who in the organization constantly sees or hears (monitors) what is going on?
>
> Who has seen the most near-misses, or heard the most scare stories—stories that are often seen as humorous, but contain implications of tragedy?
>
> Who in the organization has the most intimate knowledge of the things that make work more difficult, that over time cause discomfort, pain, or stress?
>
> Who in the organization has the best "feel" for management messages that are perceived as conflicting and force choices to be made, e.g., output expectations versus safety expectations?
>
> Who in the organization has the best "ear" to learn how people feel about the safety program, especially its weaknesses?
>
> And, who has the most to gain personally from continual improvement in the system of controlling the results of unwanted energy transfers or environmental conditions?

The answers to these questions reveal that both management and the workers have vested interests in worker involvement in a quality safety program, a program that is in a state of continual improvement. Management has a vested interest in the information the workers can provide, and in the suggestions they can make, and workers have a vested interest in the benefits of improved safety conditions. Active worker involvement in the administration of the barrier systems for controlling the results of unwanted energy transfers or environmental conditions is a "win-win" situation. Not only that, it is also an essential element of Total Quality Safety Management.

7

The Strategies of Prevention by Safety Program Systems

Figure 5–1, "The Anatomy of an Accident," identifies the nature of the work of safety by establishing the differences between safety tactics and safety strategies. Safety tactics, the use of barrier systems, are primarily directed at controlling the results of the hazard related incident (HAZRIN). Safety strategies, the use of safety program systems, are directed at the prevention of the HAZRIN, to improvement of the systems that produced the HAZRIN.

In the MORT there are four tactical systems of barriers. The strategic programs which constitute the systems for preventing incidents are far greater in number and complexity. However, within the MORT diagram, the "Land of What Happened," the safety program systems can be organized into six basic systems. These are presented in Figure 7–1.

In Figure 7–1 two departures in terminology from the original MORT analytical logic diagram are made. The term "Supervision LTA" used in the original analysis of specific control factors is renamed "Operations Management LTA". This was done to depersonalize the title and emphasize the functional nature of the category. Attention must concentrate on what is being done or is not being done, and not diverted to who is or is not doing it. Operations management should be understood to include middle managers, supervisors and staff services to operations.

The second departure is related to the first. "Higher Supervision Services LTA" has been transferred out of the "What Happened?" path. At this point the interdependencies between "Operations Management" in the "What Happened?" path and

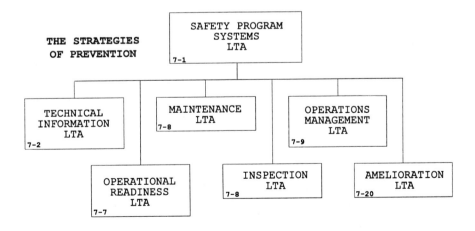

Figure 7–1. Diagram of the basic Safety Program Systems.

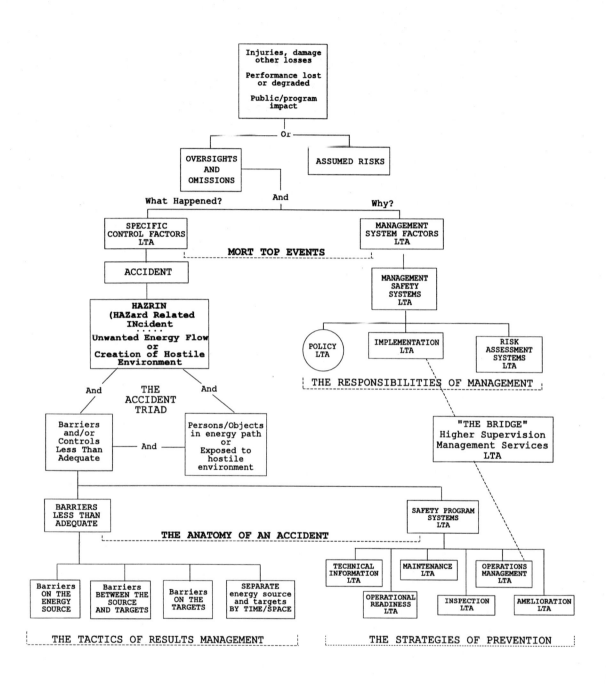

Figure 7–2. The Management Oversight Risk Tree (MORT) restated.

"Management Safety Systems" in the "Why?" path can be seen as operative. In fact, "Higher Supervision Services" can be viewed as the bridge that connects the "Land of What Happened?" with the "Land of Why?" and vice versa. This revision enables construction of a restated master diagram of the Management Oversight Risk Tree, as shown in Figure 7–2.

In Chapter 4, it was commented that the MORT Analytical Logic Diagram can be viewed as a picture of an organization's safety culture. As such it presents the viewer with a complexity that borders on being overwhelming. However the diagram is in fact a compilation of interwoven safety management systems. The need is to pull the individual systems out of the master diagram to create smaller pictures that lie within the overall fabric. Cultural changes are rarely cataclysmic. They occur as the result of small changes that occur continuously over time. This is precisely the challenge that is presented by Total Quality Safety Management; incremental changes in safety systems that result in continual improvement in the quality and productivity of our work.

It was also noted in Chapter 4 that the MORT is stated in the past tense. It is a look back at a HAZRIN: what happened and why? However, for analytical purposes, it is not necessary to wait for another HAZRIN to occur. Translation from the past tense to the present tense is all that is required. Johnson has pointed out that increased understanding of the MORT will come from analyzing a serious accident or from continuing evaluation of the safety program. The MORT is intended to be used as both a reactive and proactive analytical tool.

Figure 7–3 presents the basic graphic symbols and legends used in the diagrams. In these diagrams the event will occur if any one or more of the inputs occur. In those few situations where all of the inputs must occur to produce the event the word "and" is inserted into the diagram.

At this point we can examine the major safety program systems that have been identified above using the Less Than Adequate (LTA) terminology of the MORT. This will be done by presenting adapted versions of the MORT analytical diagrams and the respective guidance questions as presented in the *MORT User's Manual*. [Editor's Note: Excerpts from the *MORT User's Manual*—with minor adaptations—appear in italics throughout this chapter.]

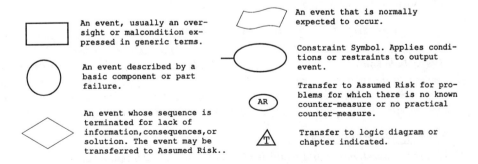

Figure 7–3. Basic graphic symbols and legends used in the MORT diagrams.

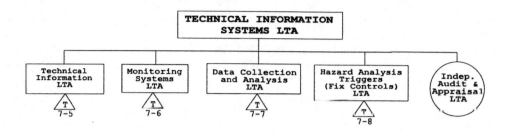

Figure 7-4. Technical Information Systems LTA.

Technical Information Systems LTA (Figure 7-4)

The *MORT User's Manual* prefaces the Technical Information Systems discussion with the following question and comment:

With respect to the unwanted energy flow [the HAZRIN], *was the technical information system adequate?*

Complex work flow processes must be supported by complete technical information systems. It is axiomatic that complex systems will depart from plans and procedures to some degree. Therefore, information systems need to detect deviations, determine rates and trends, initiate corrections, and in general, ensure that goals are attained. MORT conceives a technical information system as consisting of "research" persons, and "action" persons obtaining, handling, and providing technical information relevant to the work flow communications "network".

Technical Information LTA (Figure 7-5)

Was there adequate technical information relevant to the work flow process?

Often relevant information exists but it is not available to the action persons associated with the process. Possible reasons are investigated by the following series of questions.

KNOWLEDGE LTA

Was knowledge of the work flow process adequate? (The question is investigated by subdividing into known and unknown precedents.)

KNOWN PRECEDENT (FOR THE PREVENTION OF THE UNWANTED ENERGY FLOW)

Was application of knowledge obtainable from codes, manuals, etc., adequate? Was the list of experts to contact for knowledge adequate? Was any existing but unwritten precedent relevant to the work flow process (part of the supervisor's regular practice) known to the action person? Were there studies directed to the solution of known work flow problems? Was the effort being spent in the search for their solution reasonable and adequate?

NO KNOWN PRECEDENT

Was there investigation and analysis of prior similar accidents/incidents of the work flow process accident potential? Was the investigation adequate? Was there research directed to the obtaining of knowledge about the work flow process? Was the research effort reasonable and adequate?

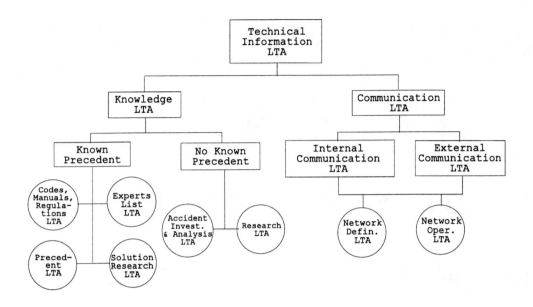

Figure 7–5. Technical Information LTA.

COMMUNICATION LTA

Was the exchange or transmittal of knowledge adequate in proportion to the potential for unwanted energy flow?

INTERNAL COMMUNICATION LTA

Was the internal communication network adequately defined? Did the internal communication network operate adequately?

EXTERNAL COMMUNICATION LTA

(These questions relate to the interface between in–house information systems and external information systems such as safety councils, professional and trade associations, government sources, etc.)

*Was the external communication network adequately defined? Did the external communication network operate adequately? Was the method of searching, retrieving, and processing **relevant** information adequate?*

Monitoring Systems LTA (Figure 7–6)

Highly complex work flow processes require excellence from the technical information system's monitoring subsystem. Triggers for fast action fixes and data for long–range hazard reduction goals are *generated* by the monitoring systems and *transmitted* by the technical information systems for managers to use in hazard analysis and risk assessment.

Was the monitoring system adequate? Were the principle elements of a good monitoring system present?

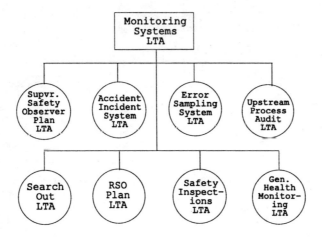

Figure 7–6. Monitoring Systems LTA.

SAFETY OBSERVER PLAN LTA

Was the safety observation plan employed by work flow process supervisors adequate?

SEARCH OUT LTA

Was there a planned independent search out effort by a safety professional for high potential hazards? Was the safety inspection search out effort adequate?

ACCIDENT INCIDENT SYSTEM LTA

Was information about incidents and accidents recorded and reviewed, and analyzed in the light of incidents and accidents that occurred in similar processes?

REPORTED SIGNIFICANT OBSERVATION (RSO) SYSTEM LTA

Was there a planned reported significant observation (RSO) system? Was the RSO system operative? (The RSO concept relates to the study of near–miss injury incidents and significant interruptive incidents observed and reported by line supervisors and workers.)

DISCREPANCY (ERROR) SAMPLING SYSTEM LTA

Was there an error sampling plan? Was it operating properly? The title, Error Sampling Plan, can evoke a prejudgment when interpreted as "Worker Mistake Sampling Plan." Assigning mistakes to workers can be done only after investigating and analyzing the incident. The title, Discrepancy Sampling Plan, is an awkward expression, but it makes no prejudgment.

SAFETY INSPECTIONS LTA

Were the routine work site safety inspections made? Were they adequate?

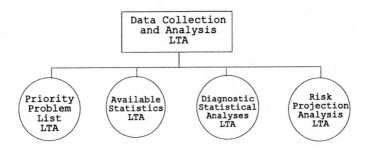

Figure 7–7. Data Collection and Analysis LTA.

UPSTREAM PROCESS AUDIT LTA

Was the audit of "upstream" work flow processes conducted adequately? (MORT separates the general work flow process into: (1) **work site operations,** and (2) **upstream work flow processes,** such as design, construction, selection, worker training, etc. Each segment must be examined in the light of the three basic work ingredients: hardware, procedures, and people.)

GENERAL HEALTH MONITORING LTA

Was the general health monitoring of the work flow process personnel adequate?

Data Collection and Analysis LTA (Figure 7–7)

Were the data collection and analysis procedures adequate? Were there analyses (i.e., measurement techniques) made of the data? Did the analyses provide the proper risk assessment information for the decision maker responsible for the risk assumption?

PRIORITY PROBLEM LIST LTA

Was there a priority problem list? Was the list kept current? (Management should know at all times what its most significant assumed risks are. Any delay in corrective action becomes an assumed risk for the present.)

AVAILABLE STATISTICS LTA

Were the available status and predictive statistics adequate?

DIAGNOSTIC STATISTICAL ANALYSES LTA

Were statistical methods (for example, flow charts, run charts, Pareto charts, cause–and–effect charts, histograms, and correlation charts) *adequately employed to identify, analyze, prioritize, and solve problems?*

RISK PROJECTION ANALYSIS LTA

Was the risk projection analysis adequate?

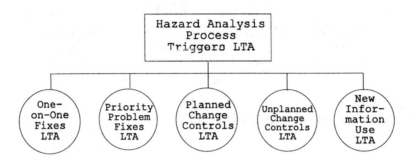

Figure 7–8. Hazard Analysis Process Triggers LTA.

Hazard Analysis Process Triggers LTA (Figure 7–8)

Were triggers (stimuli) for the initiation of the Hazard Analysis Process (HAP) adequate? Were they used to obtain early safety participation and review in planned or unplanned changes?

(MORT postulates HAP triggers as part of the HAP portion of the Risk Assessment System of Management Safety Systems, but originating from the Monitoring Subsystem of the Technical Information System within the Safety Program Systems.)

ONE-ON-ONE FIXES LTA

Was the information adequate to trigger one–on–one preventive action (for example, fix a particular piece of equipment)?

PRIORITY PROBLEM FIXES LTA

Were problems being analyzed, prioritized, and acted upon?

PLANNED CHANGE CONTROLS LTA

Were planned changes in work processes adequately recognized? Were they used to initiate safety analysis?

UNPLANNED CHANGE CONTROLS LTA

Were unplanned changes that were detected in the work process adequately recognized? Did they prompt safety analysis?

NEW INFORMATION USE LTA

Was new information from research, new standards, regulations, etc., detected and used?

Independent Audit and Appraisal LTA

Was there a recent appraisal of the total safety system (or audits of parts thereof)? Were the audits conducted in a truly independent manner? Was the appraisal plan adequate?

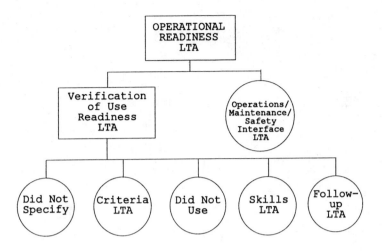

Figure 7–9. Operational Readiness LTA.

Operational Readiness LTA (Figure 7–9)

Was the facility and process operationally ready? Were the necessary supplementary operations supportive to the main process ready?

This branch supports the status of "upstream processes" (design, training, etc.) which supports the ingredients of the work process (hardware, procedures and people). The ingredients used at the worksite are obtained from two major upstream processes: (1) the original design, construction, test, and qualification, plus documents defining operation limits and performance specification, and (2) modification projects to the facility. All "upstream processes", including the Hazard Analysis Process, are susceptible to constructive analysis as "work processes" in themselves. Each upstream process can be analyzed as to hardware, procedures, and personnel.

(This MORT discussion of Operational Readiness concentrates on a new facility, or post major change start-up. However, premature start-up of equipment that has been out of service for adjustment, minor maintenance, etc., can be the source of trouble.)

Verification of Use Readiness LTA

Was verification of the facility, and/or equipment or work process adequate?

DID NOT SPECIFY
Was the conduct of an operational readiness check specified?

CRITERIA LTA
Was the criteria for determining readiness adequate?

DID NOT USE
Was the required check procedure followed?

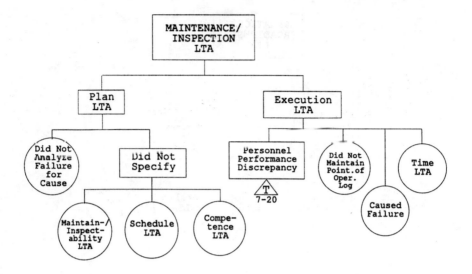

Figure 7–10. Maintenance/Inspection LTA.

SKILLS LTA
Were the personnel who made the decision on use readiness skilled and experienced?

FOLLOW-UP ACTION LTA
Were all outstanding action items resolved prior to the start up of the work flow process?

Operations/Maintenance/Safety Interface LTA

Were the interface relationships between Operations, Maintenance, and Safety adequate?

Maintenance/Inspection LTA (Figure 7–10)

Was there adequate maintenance (or inspection) of equipment, processes, utilities, operations, etc.? The analytical logic in Figure 7–10 applies to both the Maintenance and Inspection program systems.

Plan LTA

Was the scope of the plan broad enough to cover all areas that should be maintained (or inspected)? Was management aware of those areas not included in the plan?

DID NOT ANALYZE FAILURE FOR CAUSE
Did the plan require that any failed item be analyzed for the cause of failure? Was action by the appropriate individual or group required on the analysis results?

DID NOT SPECIFY

MAINTAINABILITY (OR INSPECTABILITY) LTA

Did the plan address methods for minimizing problems with equipment, processes, utilities, operations, etc., when they were undergoing maintenance (or being inspected)?

SCHEDULE LTA

Was there a schedule? Did the plan schedule maintenance (inspections) frequently enough to prevent or detect undesired changes? Was the schedule readily available to maintenance (inspection) personnel? Was the schedule coordinated with operations to minimize conflicts?

COMPETENCE LTA

Did the plan specify minimum requirements for competence and training of individuals assigned to do the work?

Execution LTA

Was the execution of the maintenance (or inspection) plan adequate?

PERSONNEL PERFORMANCE LTA

Were the tasks (as set forth by the plan) performed properly? Here the analytical logic transfers to the Personnel Performance Discrepancy lower tier event under Operations Management Systems (Figures 7–11 and 7–13).

DID NOT MAINTAIN A "POINT OF OPERATION" LOG

Was there a log of maintenance (inspections) kept at the point of operation on the piece of equipment, process, etc.? (This is distinct from other logs that may be kept in a control room, back at the main office, or in someone's desk or file. Familiar examples would be the inspection tags found on fire extinguishers, or the inspection notice in elevators.)

CAUSED FAILURE

Was the maintenance (inspection) performed without itself causing a failure or degradation of the process?

TIME LTA

Was the time specified in the plan's schedule sufficient to adequately perform the task at each station? Was the time budgeted for personnel adequate to fulfill the schedule? Was the time actually provided? Who verifies that the schedule is being met?

Operations Management LTA (Figure 7–11)

Was worksite supervision adequate? Were the necessary supportive services adequate?

In discussing Figure 7–1, it was pointed out that in this adapted version of the MORT diagram and *User's Manual*, "Supervision LTA" was renamed "Operations Management LTA," to emphasize the category's functional nature. The category includes middle operations managers and staff services to first line supervisors.

In this diagram, the MORT category "Task Performance Errors" has been renamed "Task Performance Discrepancy." This change has also been made to emphasize the logic involved in seeking system improvement. The phrase is awkward, but we need to neutralize everyone's understanding of the concept of "Error."

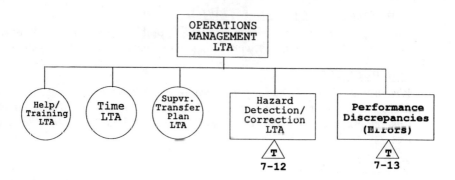

Figure 7–11. Operations Management LTA.

MORT identifies the first line supervisor as a "key person" in worksite safety. If supervisors are to adequately fulfill their responsibilities, they must have competent and useful advice and support from several kinds of supportive services. The adequacy of site supervision is therefore examined by MORT in this broader context, and tries to assess management's role in support and service to the supervisor. The emphasis throughout is to discuss what in the management system failed—not who.

Help and Training LTA

Were the help and assistance given supervisors adequate to enable them to fulfill their roles? Was the feedback of information to the supervisor adequate? Was it furnished in a usable form?

What training had the supervisor been given in general supervision? What training had the supervisor been given in safety? Has the supervisor training program been evaluated?

Time LTA

Did the supervisor have time to thoroughly examine the job (to develop an awareness of what was going on)?

Supervisor Transfer Plan LTA

Were there any gaps or overlaps in supervisory assignments related to the event? If the supervisor was recently transferred to the job, was there protocol for orderly transfer of safety information from the old to the new supervisor?

Hazard Detection/Correction LTA (Figure 7–12)

Were adequate efforts made to detect/correct hazards?
Knowledge of hazards is available from the work force. Systems that actively involve workers in the processes of hazard detection and correction must be in place.

DID NOT DETECT HAZARDS

When did the supervisor last make an inspection of the area? Was any unsafe condition present in this accident/incident also present at the time of the inspection? Was the condition detected?

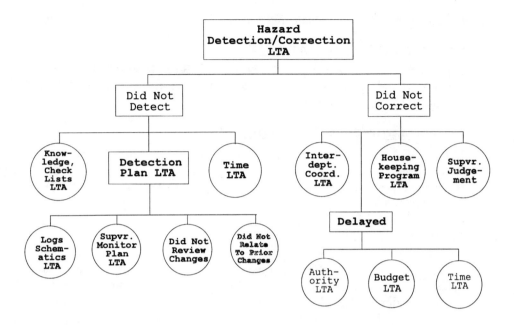

Figure 7–12. Hazard Detection/Correction LTA.

KNOWLEDGE/CHECKLISTS LTA

Was a checklist specific to the process or area available? Was it used? Was the supervisor considered generally competent to assess safety aspects of his or her area of work?

DETECTION PLAN LTA

Was there an overall detection plan for uncovering hazardous conditions?

Logs/Schematics LTA. *Was the point–of–operation posting of warnings, emergency procedures, etc., provided for in a general detection plan? Were maintenance and inspection records adequate? Were work schematics adequate? Were equipment change tags used?*

Supervisor Monitor Plan LTA. *What guidance was given to the supervisor relative to inspecting and monitoring equipment, procedures, and personnel? Was the guidance used? Was the supervisor given guidance on detection of individual personnel problems, such as alcoholism, drug use, and personal problems?*

Did Not Review Changes. *Was guidance given on the significance of change and on review methods and change detection? Were the changes involved known to the supervisor? What counterchanges were made for the known changes?*

Did Not Relate to Prior Incidents. *If there were any known prior incidents afflicting the process was the supervisor told they might correlate with safety incidents? Had the supervisor made an effort to correlate them? Was the supervisor aware of other signs or warnings that the process was moving out of control?*

Time LTA

Did the supervisor have adequate time to detect the hazards?

DID NOT CORRECT HAZARDS

Was an effort made to correct the detected hazards? (Some facts about noncorrection of hazards were dealt with under nondetection. There are some basic factors of noncorrection still to be examined.)

Interdepartmental Coordination LTA

If the accident/incident involved two or more departments, was there sufficient and unambiguous coordination of interdepartmental activities? (Interdepartmental coordination is a key responsibility of the first line supervisor. It should not be left to work level personnel.)

Correction Delayed

Was the decision to delay correction of the hazard assumed by a supervisor of production or of maintenance, on behalf of management? Was the level of risk one the supervisor had authority to assume? Was there precedent for the supervisor assuming this level of risk (as then understood by the supervisor)?

Note that a decision to delay correction of the hazard may or may not transfer to be an assumed risk. It was an assumed risk only if it was a specific named event, analyzed, calculated where possible, evaluated, and subsequently accepted by the supervisor who was properly exercising management–delegated authority.

Authority LTA. *Was the decision to delay hazard correction made on the basis of limited authority to stop the process?*

Budget LTA. *Was the decision made because of budget considerations?*

Time LTA. *Was the decision made because of time considerations?*

Housekeeping Program LTA

Was the housekeeping, as an ongoing program, adequate? Was the storage plan for unused equipment adequate?

MORT comments that the true role of housekeeping in the accident experience is usually unclear. This may be true, but the positive role of housekeeping in safety programs has been commented on earlier in the discussion of barriers, Chapter 6.

Supervisory Judgment LTA

Was the judgment exercised by the supervisor to not correct the detected hazard adequate, considering the level of risk involved? If there were previously established supervisor authority limitations, were the supervisor's actions generally in accord with those limitations?

Evaluation of the performance of a supervisor in a given situation is, of course, retrospective and must be fairly considered. If the authority limitations have been defined (as they should be), then the adequacy of a supervisor's performance is more easily measured.

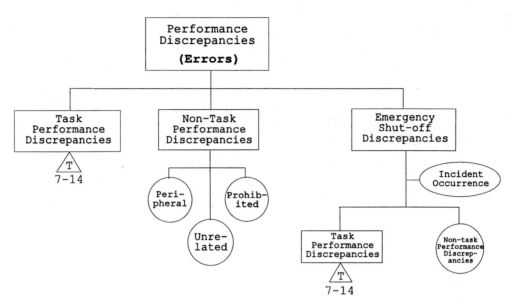

Figure 7–13. Performance Discrepancies (Errors).

Performance Discrepancies (Errors) (Figure 7–13)

Was the work activity at the worksite free of performance discrepancies (errors), by worksite personnel?

The MORT analysis separated performance errors into **task, non–task,** and **emergency shut–off errors.** Worksite activity can be viewed as usually proceeding in a normal manner to attain performance goals. If the ongoing activity enters a nonnormal phase which requires shutting off work processes, it is described as an emergency, and is analyzed in the light of the additional stress associated with emergency action. The analysis proceeds more easily with these considerations.

It should be pointed out that the kinds of questions raised by MORT are directed at systemic and procedural problems. Experience to date shows there are few "unsafe acts," in the sense of blameful work-level employee failures. Assigning the term "unsafe act" to an incident should not be done unless or until the preventive steps of (1) hazard analysis, (2) management or supervisory detection, and (3) procedures safety review have been shown to be adequate.

For the purposes of this discussion the awkward expression "discrepancy" is used to emphasize that the search is for systemic and procedural problems and to depersonalize such expressions as "unsafe act" and "error". In practice those more convenient terms will be useful only when everyone accepts them as totally neutral in meaning.

TASK PERFORMANCE DISCREPANCIES (Figure 7–14)

Was the task–related work activity free of hazards caused by performance discrepancy?

TASK ASSIGNMENT LTA

Was the task assignment of proper scope, with steps and objectives clearly defined? Was the task assignment one the supervisor should have made?

67

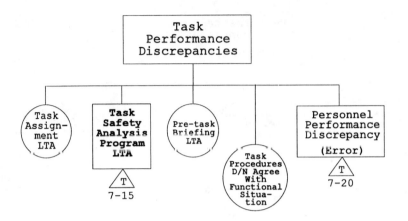

Figure 7–14. Task Performance Discrepancies.

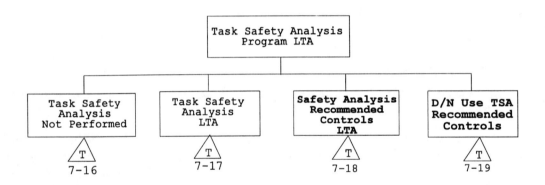

Figure 7–15. Task Safety Analysis Program LTA.

TASK SAFETY ANALYSIS PROGRAM LTA (Figure 7–15)

The standard criterion of adequacy used in MORT is Job Safety Analysis (JSA). JSA is defined as a written step–by–step procedure prepared by supervisors and/or craftsmen and approved by operations management and safety. The selection of this level of excellence is based on the proven effectiveness of JSA. It is a universal requirement in large numbers of well–run organizations.

Operating procedures may be as good as JSA or better if they have had shop floor review and input. Both may be applicable (for example, an engineering procedure for changing out a major piece of equipment may be amplified by a crane operation covered by JSA).

For this discussion the expression "Task Safety Analysis" has been employed to more narrowly proscribe the content of the resulting document.

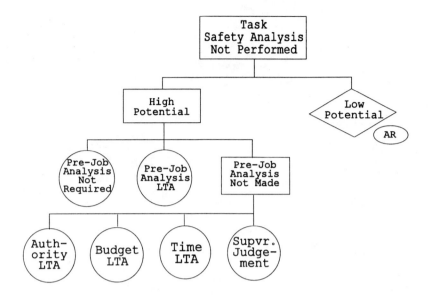

Figure 7–16. Task Safety Analysis Not Performed.

Task Safety Analysis Not Performed (Figure 7–16). *Was any form of task safety analysis performed as part of the work process?*

Effort directed to task safety analysis should be scaled to fit the magnitude of the safety hazard posed by the work task. The effort applied to work processes having high energy potential is usually highly formalized. The analysis results are implemented by a written procedure developed by the task supervisor and a small group of the most skilled craftsmen, and will usually be subjected to independent review. At the other end of the spectrum is the informal, oral review by the task supervisor before work level personnel start to work. This latter level of safety analysis is applied to tasks having low energy or low hazard potential. It is used most often with tasks related to routine maintenance and repair activity and will usually not have been independently reviewed.

The task safety analysis level of effort actually applied will range somewhere between the extremes described. MORT uses the concept of Pre–job Analysis, by which is meant that nearly every task must be surveyed step by step to determine the level of effort of task safety analysis to be performed. The MORT diagram analysis proceeds with the premise that pre–job analysis should **always** be made for tasks assessed as having significantly high hazard potential.

High Potential
> *Was an analysis performed for a work task involving a high potential for discrepancy, injury, damage, or for encountering an unwanted energy flow?*

> Pre–Job Analysis Not Required
>> *Did the operations management require a pre–job analysis to the scale of the magnitude of the task safety analysis to be performed?*

> Pre–Job Analysis LTA
>> *If required, was the pre–job analysis, as performed, adequate to the scale of the magnitude of the task safety analysis to be performed?*

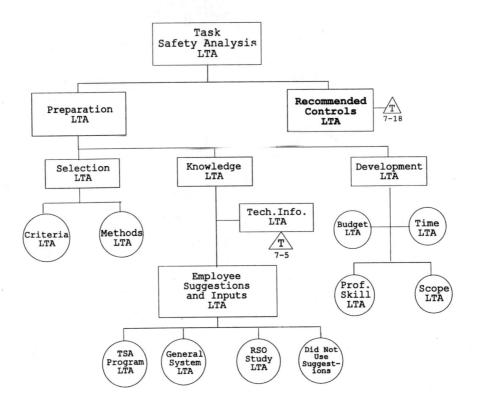

Figure 7–17. Task Safety Analysis LTA.

        Pre–Job Analysis Not Made
            Was a pre–job analysis required but not made? Was it not made because of
- lack of authority?
- budget reasons?
- schedules?
- a decision by a line supervisor?

    Low Potential
        Was the work task assessed as one involving low potential? Was this a reasonable assessment? Was the decision not to perform a task safety analysis properly delegated to the supervisor?

Task Safety Analysis LTA (Figure 7–17). Was the task safety analysis adequate? Was it properly scaled for the hazards involved?

    Preparation LTA
        Was the preparation and content of the task safety analysis adequate?

        Selection LTA
            Were the hazards associated with the work adequately identified and selected? Were the criteria used adequate? Were the methods used in prioritizing the identified hazards adequate?

Knowledge LTA

Was the knowledge input to the task safety analysis adequate?

Employee Suggestions and Inputs LTA

Was consideration of employee–developed suggestions and inputs adequate?

Team Program LTA: *Was a team used to obtain work-level employee participation? Was the process of accomplishing the task safety analysis (TSA) program adequately defined and staffed?*

General System LTA: *Was the general system for collecting and utilizing other employee suggestions and inputs adequate?*

RSO Study LTA: *Were Reported Significant Observation (RSO) studies used to gather employee inputs? Were these RSOs readily accessed?*

Did Not Use Suggestions: *Were employee suggestions and inputs made through other processes used?*

Technical Information LTA

Was the technical information (with respect to the preparation of the safety analysis) adequate? (Technical information relevant to safety aspects of the work processes often exists but is not available to the "action" persons associated with the process. Possible reasons for this are investigated by transfer to Figure 7–3.)

Development LTA

Was the development of the specific TSA by the first line supervisor adequate? If judged to have been inadequate, what were the true underlying causes of the inadequacy? An honest assessment should be made of what could reasonably be expected of the supervisor, taking into account existing time and budget restrictions placed on him or her by higher supervision.

Time LTA

Was there time for an adequate development of task safety analysis?

Budget LTA

Were there sufficient departmental funds?

Scope LTA

Were the scope and depth of the task safety analysis development sufficient to cover all related hazards?

Professional Skill LTA

Were the experience and skill of the supervisor and other participants adequate to accomplish the required work task safety analysis?

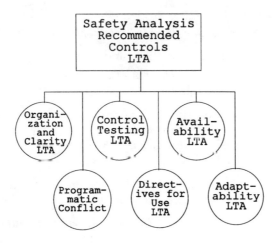

Figure 7–18. Safety Analysis Recommended Controls LTA.

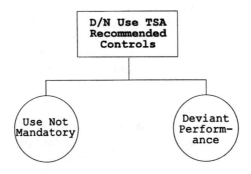

Figure 7–19. Did Not Use Recommended Safety Analysis Controls.

Safety Analysis Recommended Controls LTA (Figure 7–18). *Were adequate worksite controls placed on the work process, facility equipment, and personnel by the task safety analysis?*

    Organization and Clarity LTA
        Were the organization and clarity of presentation of the task safety analysis recommended controls adequate to permit their easy understanding and use?

    Programmatic Conflict
        Were the recommended controls free of conflict with the overall project goals and requirements?

    Control Testing LTA
        Were recommended controls tested at the worksite for feasibility before being directed for use?

Directive For Use LTA
> *Was the management directive for use of the task safety analysis recommended controls adequate? Was it explicit and not subject to possible misunderstanding?*

Availability LTA
> *Did the management information system make knowledge of the recommended controls available to worksite personnel?*

Adaptability LTA
> *Were the recommended safety controls made in a form which allowed them to be adequately adapted to the varying conditions?*

Did Not Use Safety Analysis Recommended Controls (Figure 7–19). *Were the safety controls recommended by the task safety analysis being used?*

Use Not Mandatory
> *Was use of the recommended safety controls mandatory? (If use was not mandatory, failure to use them is either an assumed risk or a management safety system failure.)*

Deviant Performance
> *If use of the recommended safety controls was mandatory, were they actually used? (If use was mandatory, failure to use them is a deviant performance on the part of the line supervisor.)*

Pre–Task Briefing LTA

Was the work force given a pre–task briefing (prior to task performance)? Was it adequate? Did the pre–task briefing adequately consider the net effect of recent changes, maintenance, new hazards, etc.? (See Figure 7–14.)

Task Procedure Did Not Agree with Functional Situation

Did the work task completion procedure, as directed by oral or written instructions, agree with the actual requirements of the work task? (Direction or requirements, as defined by specifications, operating procedures, equipment manuals, etc., may conflict with actual work task requirements.) (See Figure 7–14.)

Personnel Performance Discrepancy (Error) (Figure 7–20)

Did the individuals assigned to the work perform their individual task assignments properly? (Possible causes of performance discrepancy should be considered for each individual whose performance was judged to be discrepant.)

The discrepancy rate in the ongoing process is not a function of the intellect. It is a function of the design of the system.

Personnel Selection LTA: *Were personnel selection methods adequate?*

Criteria LTA
> *Were the safety–related requirements of the job adequately defined so as to select an individual with desired characteristics?*

Testing LTA
> *Did the individual meet the standards established for the task? Had the assigned individual been recently re–examined to the standards established for the task?*

Training LTA: *Was the training of personnel adequate?*

None
> *Was the individual trained for the task he or she performed?*

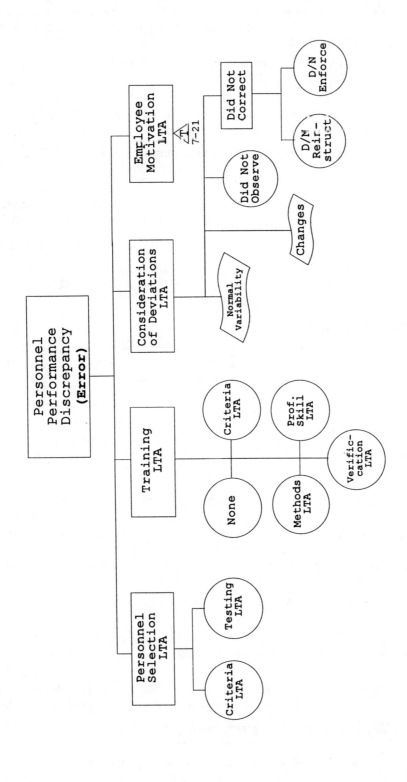

Figure 7–20. Personnel Performance Discrepancy (Error).

Criteria LTR
: *Was the criteria used to establish the training program adequate in scope, depth, and detail?*

Methods LTA
: *Were the methods used in training adequate to the training requirements? (Consider methods such as realistic simulation, programmed self–instruction, and other special training in addition to basic instruction, plant familiarization, etc.)*

Professional Skills LTA
: *Was the basic professional skill of the trainers adequate to implement the prescribed training program?*

Verification LTA
: *Was the verification of the person's current trained status adequate? Were retraining and requalification requirements of the task defined and enforced?*

Consideration of Deviations LTA: *Was the verification shown by the supervisor for the need to observe deviant personnel performance?*

The analysis shows contributions to deviations from both normal variability and changes. Normal personnel performance variability is viewed as manageable through appropriate equipment design, ergonomics, good planning, training, and application of human factors. Change is characteristic of illness, fatigue, personal problems, etc., which results in individual performance outside the normal range variability.

Normal Variability
: *Was the deviation in personnel performance within the range of normal variability? (The banner event symbol is used to show that some degree of variability is normal and expected.)*

Changes
: *Was the deviation in personnel performance significantly different than the performance standard for the task? (Again, the event symbol is used to show that some degree change is normally expected to occur.)*

Did Not Observe
: *Was the deviation (i.e., extreme variability or significant change) observed by the line supervisor?*

Did Not Correct
: *Did the supervisor act to correct the observed personnel performance discrepancy?*

> Did Not Reinstruct
> : *Did the supervisor reinstruct the person observed as to the correct procedure?*
>
> Did Not Enforce
> : *Did the supervisor enforce the established correct rules and procedures? Were disciplinary measures taken for willful and habitual disregard of the rules and procedures?*

Employee Motivation LTA (Figure 7–21). *Were the employee motivation, participation and acceptance adequate?*

Employee motivation plays a significant role in personnel performance in accomplishment of the work task. Various aspects of employee motivation are analyzed by lower tier events.

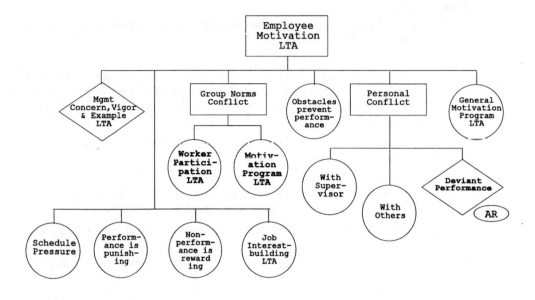

Figure 7–21. Employee Motivation LTA.

It is at this precise point in the MORT Analytical Logic Diagram that the abandonment of the command–control theory of organization management for the new paradigm of consensus management impacts most directly and with greatest force. The new paradigm does not change the nature of the questions or areas of concern, but it significantly affects the methods necessary for resolving the questions. The questions raised remain pertinent and valid—it is the answers that will present themselves as revolutionary.

> Management Concern, Vigor, and Example LTA
>> *Was management concern for safety displayed by direct vigorous personal action on the part of top executives?*
>
>> In TQSM the objective of the safety program is the elimination of waste: the waste of assets, the waste of materials, and the waste of time at all levels that arise out of accidents. This imperative directly serves the economic viability of the organization and will command concern, vigor, and example at all levels of management. It could be added that the vigorous personal actions must be persistent.
>
> Schedule Pressure
>> *Were task schedule pressures (as experienced by the individual) held to acceptable levels? Were there output expectations that in the employee's mind superseded safety compliance expectations?*
>
> Performance is Punishing
>> *Was the employee fairly treated for performing as supervision desired? (From the viewpoint of the employee, sometimes there is an undesirable consequence for doing a good job.)*

Non-Performance Is Rewarding

Did the employee find the consequence of doing the job incorrectly more favorable than doing the job as directed? (Obstructive behavior may be more rewarding to the individual.)

Job Interest Building LTA

Does performing the task well really matter to the individual performing it?

Under traditional management the performing individual can often believe the consequence is the same whether the task is done right or some other way.

Group Norms Conflict

Are the actions and attitudes of the individual's peer groups in harmony with the task requirements and the goals of the larger organization?

Worker Participation LTA

Was there adequate opportunity for the worker to participate in analysis, training, or monitoring systems?

Motivation Programs LTA

Was there adequate use of management motivational programs to develop desired behavioral change in individuals?

Obstacles Prevent Performance

Were obstacles that might prevent task performance reduced to an acceptable level?

Personal Conflict

Are individual conflicts, which may have a negative relationship to task safety, adequately resolved in the individual? Does the individual have good standards of judgment?

With Supervisor

Were employee and supervisor personalities compatible in the work place?

With Others

Was the employee compatible with other workers in the work place?

Deviant Performance

Were the psychological traits of the individual judged acceptable when rated against the task safety requirements? (Individuals with abnormally high levels of social maladjustment, emotional instability, and conflict with authority produce more than their share of accidents. The decision to continue an individual in a task, rather than transfer, can be an assumed risk if the energy levels are low.)

General Motivation Program LTA

Was there a general motivation program on safety, employed by management, to perform correctly? This question is redundant in quality managed organizations.

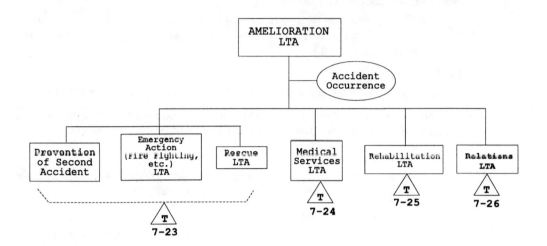

Figure 7–22. Amelioration LTA.

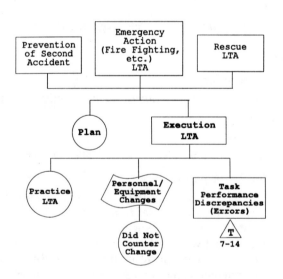

Figure 7–23. Prevention of Second Accident, Emergency Action, Rescue LTA.

Amelioration LTA (Figure 7–22)

Once an accident has occurred, was there adequate amelioration on the part of all concerned?

Amelioration can be considered and evaluated only after an accident, or with the presumption that an accident has occurred. This is indicated by the presence of the "Accident Occurrence" constraint on the gate leading to lower branch elements of the analytical logic diagram (Figure 7– 20). The intent of amelioration is to limit the consequences of an accident occurrence to **reduce the sensitivity** of those consequences whenever possible.

When evaluating amelioration from an overall management system standpoint, consider the following:
- *Are all of the amelioration functions preplanned (as opposed to having them occur fortuitously at the time of a particular accident)?*
- *Does the plan adequately scope the types and severity of accidents that it intends to cover?*
- *Are adequate resources allocated to properly execute the plan?*
- *Is management aware of any residual risk **beyond the scope of the plan?***

Prevention of Second Accident LTA (Figure 7-23)

Through the efforts of individuals at the accident scene and those who arrived later, were steps taken to prevent a second accident caused directly or indirectly as a consequence of the first?

Emergency Action (fire fighting, etc.) LTA (Figure 7-23)

Was the emergency action prompt and adequate to the emergency? Which emergency response teams were required? Were they notified and did they respond? (Include local facility fire brigade, health physics team, fire department, spill containment team, and other specialty teams. Consider delays or problems in both notification and response.)

Rescue LTA (Figure 7-23)

Were trapped or immobilized victims satisfactorily removed to a safe area? Before entering a hazardous area, did rescuers consider the risk of injury to themselves versus the ability to lessen the severity of injuries to the victims? Include the evacuation of employees or the public from potentially hazardous areas.

PLAN LTA
If properly executed, was the plan adequate to accomplish the intended function? Was the plan provided to those who needed it?

EXECUTION LTA
Was the plan executed as intended?

PRACTICE LTA
Was there sufficient practice of various plan assignments? Was the practice realistic?

PERSONNEL AND/OR EQUIPMENT CHANGES
Were there personnel or equipment changes that caused the execution of the plan to be less than adequate? Were trained personnel free of any recent physical or mental changes? Was the equipment familiar to the users and free of defects or modifications?

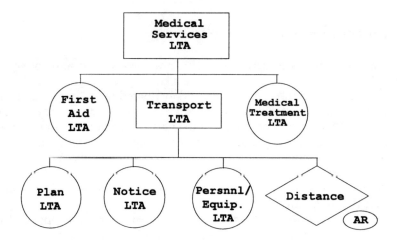

Figure 7–24. Medical Services LTA.

Did Not Counterchange. *Had appropriate counter changes been considered and introduced where applicable for changes in personnel or equipment?*

Task Performance Discrepancies (Errors)

Was the plan executed through completion of all steps? Note the transfer to lower tier events of Task Performance Discrepancies, Figure 7–12.

Medical Services LTA (Figure 7–24)

Was adequate medical service available?

First Aid LTA

Was adequate first aid immediately available at the scene? Was it used properly to prevent immediate injuries from becoming more severe?

Transport LTA

Was mobile service available to transport medical personnel and equipment to the accident scene and/or to transport injured to medical facilities? Was transport executed properly?

Plan LTA

Was there a medical service plan? Was it distributed to appropriate personnel? (Consider such things as: (1) how to make a notification, (2) training of medical personnel and drivers and knowing when they are available, and (3) who and what equipment will respond.)

Notice LTA

Was notification made in an adequate time and manner? Were employees instructed on how to notify medical services? (Consider whether the notification process was easy to do, especially during the stress of an emergency.)

Personnel and Equipment LTA

Did the personnel use the equipment correctly? Did the equipment function properly? Did the medical and transport personnel have all the equipment necessary to properly per-

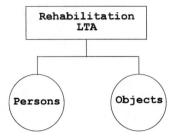

Figure 7–25. Rehabilitation LTA.

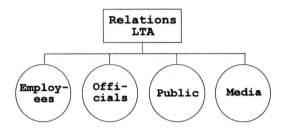

Figure 7–26. Relations LTA.

form the jobs expected of them? Were the personnel adequately trained relative to the postulated needs? (Consider whether equipment could be operated easily during the stress of an emergency.)

DISTANCE

Was there a significant distance between medical services and the area to which service responded? (If distance is great, response time is increased.) Note: The event is flagged as an assumed risk. Top management must assume distance/time response risk.

MEDICAL TREATMENT LTA

Was there adequate medical treatment enroute and at the medical facilities?

Rehabilitation LTA (Figure 7–25)

Was rehabilitation of persons and objects made after the accident?

PERSONS

If the injury was disabling, could its overall disabling effect have been reduced and/or the individual made more functional? If such rehabilitating activity was possible, was it done?

OBJECTS

Was damaged equipment, buildings, or other property expeditiously repaired, salvaged, or replaced?

81

Relations LTA (Figure 7–26)

Was there a management plan outlining the protocol to be followed and steps to be taken subsequent to a significant accident? Was the accident news disseminated to all concerned parties in a proper and timely manner?

EMPLOYEES

Did the relatives of the injured employee first hear about the accident from a responsible, tactful individual within the organization? Were the other employees in the organization notified firsthand about the accident with some assurance that significant corrective action would be taken?

OFFICIALS

Were the facts about the accident given accurately and in a timely manner to the proper officials of: (1) the organization, (2) the customer, (3) the municipality, (4) the state, and (5) OSHA and other governmental agencies as appropriate?

PUBLIC AND MEDIA

Were the news media (and thereby the public) given the accident facts and assurance that significant corrective actions were being taken? Was a specific point of contact within the organization provided as the source of additional information?

Summary Comments and "The Bridge"

The Management Oversight Risk Tree (MORT) is the continuing result of work that took place during the 1970s. Johnson's definitive book, *MORT Safety Assurance Systems*, was published in 1980. At that time American business management was just starting to rediscover the principles of management by statistical methods that had been set forth by W. Edwards Deming, Joseph M. Juran and others, and that had been employed with great success by U.S. manufacturers during the World War II production years. (This background is not mentioned in Johnson's book. And there are no expanded discussions on statistical methods in the MORT diagram and *User's Manual* when statistics are mentioned.)

The wartime experience with statistical quality control centered on the mass production of "things", materiel for the largest war in history. It was directed towards improving manufacturing systems. The values involved were finite and discrete, derived from direct measurement. When American industry returned to producing consumer goods in peacetime, it saw no need to bother with what was viewed as a complex method of quality control. After all, it had the only operational industrial base in the world. Quantity, not quality, was needed. Any use or expansion of the new methods was completely abandoned. This is familiar history.

Equally familiar is the history that developed elsewhere, most notably in Japan. The industrial start-up following the war there literally began at "ground zero". Japan had no industrial base, and to compound the problem, had the worldwide reputation of producing nothing but consumer junk. The imperatives were for both quantity and quality. In 1950, when Dr. Deming told the leaders of Japanese industry that they could take over world markets in five years by using statistical methods in the planning and operation of their industries, the leaders listened and they responded.

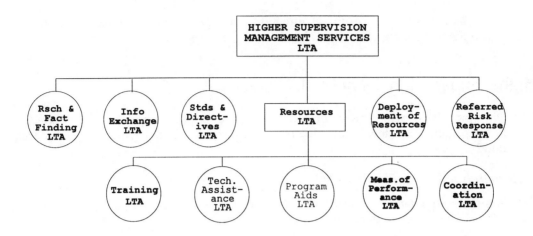

Figure 7–27. Higher Supervision Management Services LTA.

Today, as universal experience with the use of quality methods has broadened, managers are realizing that these methods work not only with the production of "things", they also apply to the provision of services. This dawning realization has grown out of the new understanding of the complex redundant nature of the activities within an enterprise.

That understanding is based on the realization that an enterprise consists of countless highly interdependent processes, or systems of activities. And, just as continuous improvement of the systems of the production of "things" results in higher and higher product quality, so too does the use of statistical methods work to continuously improve the provision of services, even in the absence of finite or discrete things to count or measure.

This view of the interdependencies of systems is not new, but only rarely has it been applied to organized human activities, although it is clearly observable in nature. In human activities we need to be able to identify these interdependent systems so attention can be directed toward their continuous improvement, system by system, while taking full account of interdependencies at the same time.

For safety, for the management of unwanted energy transfer as an organized activity, this is the basic contribution of the management oversight risk tree. The statistical method employed is the flow chart, which identifies and depicts individual small systems and combines them in logical order to create a picture of the safety culture of the organization.

"The Bridge"

Chapters 1 through 6 examined the "What Happened" factors of the accident phenomenon. Our attention can now be directed to the management side, the MORT's "Why" factors.

However, in order to arrive in the land of Management Safety Systems it is necessary to examine the "bridge" between the two different territories. Higher Supervision Services is that bridge between the Land of What Happened and the Land of Why. The analytical logic displayed in Figure 7–27 applies not only to upper line management services but to staff services as well. It is the connection between safety tactics, which control the results of unwanted energy transfers and safety strategies which prevent the transfers through safety program systems.

Higher Supervision Services LTA (Figure 7–27)

Did upper level management provide the type of support and guidance needed at lower organization levels for adequate control of unwanted work process energy flow?

Research and Fact Finding LTA

Was necessary information, which was not otherwise readily available, sought out through established research and fact finding techniques?

Information Exchange LTA

Was there an accessible open line of communication that permitted transmittal of needed information **in both directions** between upper and lower levels? Was study of a problem a shared responsibility? Were the results provided to the users?

Standards and Directives LTA

In cases where the organization and external sources of codes, standards, and regulations did not cover a particular situation, did management develop (or have developed) adequate standards and issues appropriate directions?

Resources LTA

Did management have the resources, derived from standard directives and regulations, it needed to perform the supportive services?

TRAINING LTA

Was there sufficient training to update and improve needed supervisory skills?

TECHNICAL ASSISTANCE LTA

Did supervisors have their own technical staff or access to such individuals? Was technical support of the right disciplines sufficient for the needs of supervisory programs and review functions?

PROGRAM AIDS LTA

Did management have available, for support of its programs, such aids as: useful analytical forms, training materials, reproduction services, audiovisuals, capable speakers, meeting time and rooms, technical information, monographs, etc.?

MEASURE OF PERFORMANCE LTA

Were the established methods for measuring performance that permitted the effectiveness of supervisory programs to be evaluated?

COORDINATION LTA

Were other management programs and activities coordinated with the groups and individuals who interfaced with the program participants? Did the coordination eliminate conflicts that could have reduced program effectiveness?

Deployment of Resources LTA

Were all available resources used effectively and to the greatest advantage of supervisory efforts?

Referred Risk Response LTA

Was management responsive to risks referred from lower levels? Was there an established system for analyzing and acting upon such risks in a timely manner? Was there a fast action cycle to process imminent hazard/high risk?

The Safety Responsibilities of Executive Management

Chapters 6 and 7 have presented an analysis of the Specific Control Factors on the "What Happened?" path. It is now time to turn onto the "Why?" path. The MORT identifies this path as an analysis of "Management System Factors", the Management Safety Systems that are the responsibility of executive management.

There is no subject in the practice or literature of safety that has received more consideration than the matter of "management support". In Figure 7–21, Employee Motivation LTA, mention is made of "Management Concern, Vigor, and Example LTA" without conclusion. The analysis of Management Safety Systems of the Why? path presents that conclusion. The Management Safety Systems establish the safety culture of the organization, the de facto safety policy of the organization. It has often been said that "Policy is not what is said, it is what is done", a point that is particularly relevant to Total Quality Safety Management.

8

The Safety Responsibilities of Executive Management

The concepts of risk, uncertainty and profit, and the differences between risk bearing and uncertainty bearing in the economics of the firm were discussed in Chapter 2. These key points were made:
- In business, losses can arise out of two sets of circumstances—risk, which is borne involuntarily, and uncertainty, which is undertaken voluntarily in the pursuit of profit, but with the understanding that losses are possible.
- Risk bearing is an everyday fact of life. Risk arises out of natural or human–related forces and is borne involuntarily, and can only cause loss. Risk incidents that arise out of the forces of nature are often referred to as "acts of God," while risk incidents that arise out of human–related forces are generally referred to as "accidents".
- Uncertainty bearing is the business of business, and is the true source of profit. It is where business makes its offense; it's where the fun and the high–visibility activity is and where the heroes are made. The psyche of business managers is geared to uncertainty bearing.
- Risk bearing, out of which only losses can arise, differs radically from uncertainty bearing. The prevention of losses due to involuntarily–borne risks is a business's defense, and the best that can happen is that you don't lose. It is dull, unromantic, with low visibility and no heroes; it is not much fun.

This is the landscape that surrounds the trip through the Land of Why Did It Happen—the Land of Management Safety Systems. Executive management is the architect of the structure of systems that constitutes an organization's culture. The figures in this chapter illustrate the management systems on the "Why?" side of the MORT diagram that are designed to prevent the wastes of accidents. They explore the content of what can be seen as "management support for safety". (The text for the diagrams is taken from the *MORT User's Manual*. As in Chapter 7, excerpts from the manual are italicized.)

Management Safety Systems LTA (Figure 8-1)

Are all the factors of the management system necessary, sufficient, and organized in such a manner as to assure that the overall program will be "as advertised" to the customer, to the public, to the organization itself, and to other groups as appropriate?

In the event–by–event review which follows, the questions are phrased in the present tense. Assume the diagram is being used for evaluation of an existing safety system. For accident investigation, rephrase the questions to past tense.

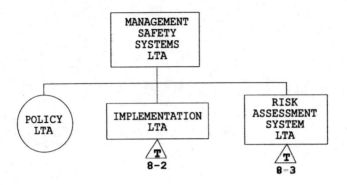

Figure 8–1. Management Safety Systems LTA.

Policy LTA

Is there a written, up–to–date policy with a broad enough scope to address major problems likely to be encountered? Is it also sufficiently comprehensive to include the major motivations (e.g., humane, cost, efficiency, legal compliance)? Can it be implemented without conflict?

Under the concepts of total quality management the sense of this question has become obsolete. One policy statement to guide the efforts of all employees and activities is all that is required. That statement will reflect the organization's dedication to customer satisfaction (both the internal and external customers), to continuous improvement of products and services, to the elimination of wastes of all types, to management by fact, to employee involvement, and to leadership in all areas.

The statement will be in terms of what everyone must be constantly working toward. The statement will, in itself, serve all motivations and eliminate conflicts through its universality. Such a statement can come only from the highest executive level. There is no need for a "safety policy statement" as such. Policy statements for separate functional groups can create conflicts and spawn adversarial relationships.

Throughout this discussion of management responsibilities, you will encounter questions reflecting traditional command/control thinking. In some cases the problem being considered would not be of great concern in quality managed organizations, with their reliance on consensus and team management methods. However, no major questions have been deleted, because the sense of the intended content does have relevance for quality management for system improvement.

Implementation LTA (Figure 8-2)

Does the overall program represent the intended fulfillment of the policy statement? If there are problems encountered in carrying out the policy, are these relayed back to the policy makers? Is the implementation a continuous, balanced effort designed to correct systemic failures, and generally proactive rather than reactive?

METHODS, CRITERIA, ANALYSES LTA

Are selective methods used for management implementation and for improving human performance? Is there a comprehensive set of criteria used for assessing the short– and long-term impact of the methods on safety for the desired results? Does management demand that adequate analyses be performed and alternative countermeasures be examined, or are criteria simplistic and therefore less than adequate?

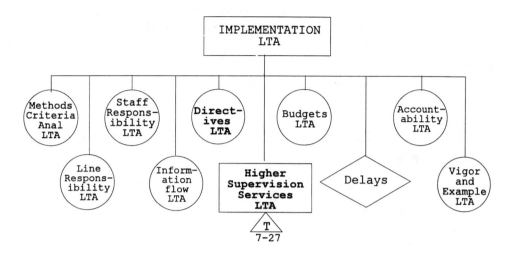

Figure 8–2. Implementation LTA.

LINE RESPONSIBILITY LTA

Is there a clear, written statement of safety responsibility of the line organization, from the top individual through the first line foreman to individual employee? Is this statement distributed and understood throughout the organization? Is it implemented?

Again, an obsolete question in a quality managed organization. What is required is that all have an understanding of the systems they are working in so that continuous improvement of those systems is a realistic responsibility.

STAFF RESPONSIBILITY LTA

Are there provisions for assigning and implementing specific safety functions to staff departments (e.g., safety, personnel and training, engineering, maintenance, purchasing, transportation, etc.)?

INFORMATION FLOW LTA

Has management specified the types of information it needs and established efficient methods by which such information is to be transmitted up through the organization? Has management, in turn, supported this process by providing the information needed at lower levels?

DIRECTIVES LTA

*Is safety policy implemented by directives which emphasize **methods** and **functions** of hazard review, monitoring, etc., rather than specific rules for kinds of hazards? Are directives published in a style conducive to understanding and without interface gaps?*

HIGHER SUPERVISION SERVICES LTA

(Refer to Figure 7–27.) Has management provided the type of supportive services and guidance needed at the lower organization levels? Is there a formal training program for all management personnel and supervision which addresses: (1) general aspects of management and supervision; (2) specific technologies; (3) human relations/communications; and (4) safety?

Note the transfer to the lower tier events in Figure 7–27. "Higher Supervision Services" is the "bridge" between "Management Safety Systems" and "Operations Management".

BUDGETS LTA

Is the budget adequate not only for the safety group but also for related safety program aspects for which other groups in the organization have responsibility?

DELAYS

Are safety program elements implemented in a timely manner? Are solutions to safety problems introduced early in the life cycle phases of projects? (Delays can and should be made known to management. If this is done and delay is a practical need, the delay becomes an assumed risk.)

ACCOUNTABILITY LTA

Is line management held accountable for safety functions under their jurisdiction? If so, are there methods for measuring their performance?

VIGOR AND EXAMPLE LTA

Have top management individuals demonstrated an interest in lower level program activities through personal involvement? Is their concern known, respected, and reflected at all management and employee levels? (Do people tell stories of a manager's vigor and example in support of continuous improvement of safety? If not, the manager's example may be LTA.)

Risk Assessment System LTA (Figure 8-3)

Does the risk assessment system provide management with the information it needs to assess residual risk and to take appropriate action if the residual risk is found unacceptable? Does the system also provide: (1) comparative evaluation of two or more systems, and (2) development and evaluation of methods supporting the hazard analysis process?

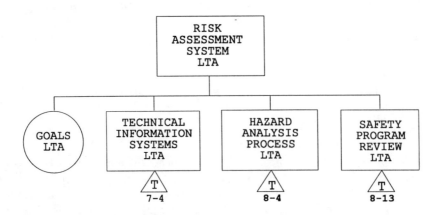

Figure 8–3. Risk Assessment System LTA.

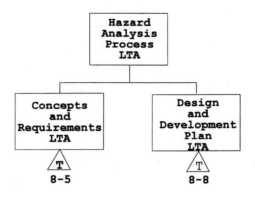

Figure 8–4. Hazard Analysis Process LTA.

GOALS LTA
Are there high goals for policy and implementation as well as specific goals for projects? Are the goals nonconflicting, sufficiently challenging, and consistent with policy and the customer's goals?

The high goals are continuous improvement of all safety systems, the elimination of wastes from accidents, and the satisfaction of safety's internal and external customers.

TECHNICAL INFORMATION SYSTEMS LTA
(Refer to Figure 7–4.) *Is the technical information system adequate to support the needs of the risk assessment system?*

Note the transfer to the lower tier events contained in Figure 7–4.

HAZARD ANALYSIS PROCESS LTA (Figure 8-4)
Is the hazard analysis process properly conceptualized, defined, and executed?

CONCEPTS AND REQUIREMENTS LTA (Figure 8-5)
Are the concepts and requirements of the Hazard Analysis Process adequately defined?

Definition of Goals and Tolerable Risks LTA. *Have goals and tolerable risks been defined for both safety and performance and any conflicts between the two resolved?*

Safety Goals and Risks Not Defined
> *Do the goals state what degree of safety excellence should be attained and when? Are tolerable direct and indirect safety risks defined and quantified?*

Performance Goals and Risks Not Defined
> *Have goals been set for performance efficiency and productivity? Have tolerable risks for lost efficiency and productivity been established and actual risks quantified? (Such goals complement safety goals by requiring greater assurance of discrepancy free performance.)*

Safety Analysis Criteria LTA (Figure 8-6). *Have the necessary criteria been specified and elements defined to adequately support the safety analysis program?*

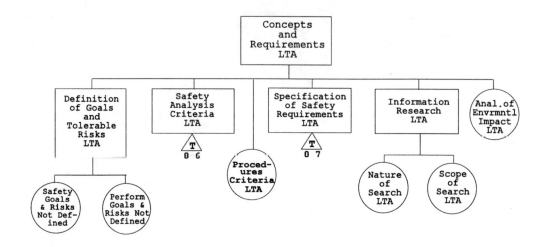

Figure 8–5. Concepts and Requirements LTA.

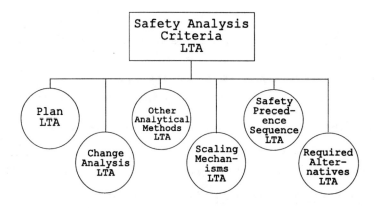

Figure 8–6. Safety Analysis Criteria LTA.

Plan LTA
: *Has a system safety plan been developed which describes "who does what and when" in analysis, study, and development?*

Change Analysis LTA
: *Has a specific change–based analytic method been established to review the form, fit, or function of components and subsystems (including interfaces) upwards in a review process, until no change is demonstrated?*

Other Analytical Methods LTA
: *Are other appropriate analytical skills available in the organization (or from a consultant) and are they used? (Examples are: Hazard Identification, Failure Modes and Effects, Fault Tree, MORT, Nertney Wheel, Failure Analysis, Human Factors Review, etc.)*

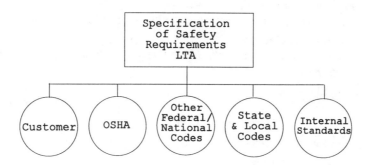

Figure 8–7. Specification of Safety Requirements LTA.

Quality management has strong preferences for the use of statistical methods in problem identification, analysis, and solution. Specific statistical methods used include: process or system flow charts, deployment flow charts, Pareto charts, Cause and Effect (Fishbone) charts, run charts, histograms, scatter or correlation diagrams, and process control charts, among others. Quality management intends that expert help and training in the application and use of statistical methods be provided all employees to the skill level required at their respective level. The original MORT question clearly predates the new methodology.

Scaling Mechanisms LTA
Has some reasonably clear–cut mechanism been established for scaling the seriousness and/or the severity of prior events? Is there a mechanism to project past events to a scaled effort to evaluate current practices?

Safety Precedence Sequence LTA
Is the preference for safety solutions prioritized as: (1) Design; (2) Safety Devices; (3) Warning Devices; (4) Human Factors Review; (5) Procedures; (6) Personnel; and (7) Acceptance of Residual Risks (after consideration of the preceding six items)?

Required Alternatives LTA
Does management require confrontation between alternative solutions in its bases for choices and decisions?

Procedures Criteria LTA. *Are engineers and designers made aware of their limitations in writing procedures for operating personnel, and of the need for selection and training criteria for operators, and of supervisory problems?*

Specification of Safety Requirements LTA (Figure 8-7). *Have all applicable and appropriate safety requirements been specified, made available, and used? Consider whether the following documents have been called out to the extent they are applicable:*

Customer Requirements
Are customer safety requirements developed in–house?

OSHA Regulations
Do OSHA regulations apply?

Other Federal and National Codes
: *By agencies other than the customer and OSHA?*

State and Local Codes
: *State and local codes applicable to the geographical area where the work is to be performed?*

Internal Standards
: *Internal standards developed within the organization to cover situations not addressed by the outside requirements?*

Information Research LTA (Figure 8–5). *Is an adequate information search required?*

Nature of Search LTA
: *Does the nature of the search include incident files; codes, standards, and regulations; change and counterchange data; related previous analyses; and other comments and suggestions?*

Scope of Search LTA
: *Is the search of sufficient scope so as to seek information on problems from conceptual design, through construction and use, to final disposal?*

DESIGN AND DEVELOPMENT PLAN LTA (Figure 8-8)

Does the development phase provide for the use of the major safety results of the "Concepts and Requirements" phase?

Energy Control Procedures LTA (Figure 8-9). *Is there an attempt, whether by design or procedure, to control energy to only the levels needed for the operation and to contain its interactions to the intended functions?*

The concepts in Figure 8–9 are discussed in depth in Chapter 6, The Concepts of Energy and Barriers.

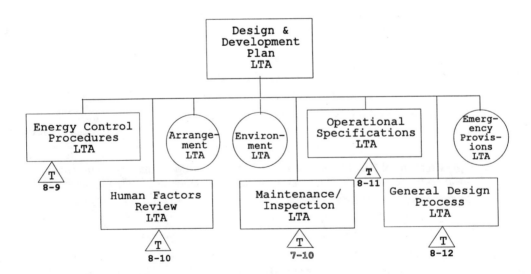

Figure 8–8. Design and Development Plan LTA.

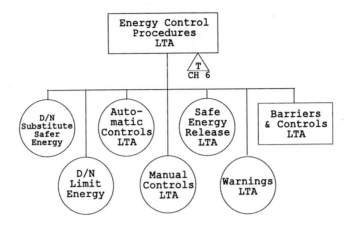

Figure 8–9. Energy Control Procedures LTA.

Human Factors Review LTA (Figure 8-10). *Have human characteristics been considered in designs, plans, and procedures as they compete and interface with machine and environmental characteristics?*

Professional Skills LTA
Is the minimum level of human factors capability needed for an operation available? Is it used?

Did Not Describe Tasks
For each step of a task, is the operator told when to begin, how to perform the step, when the step is finished, and what to do next?

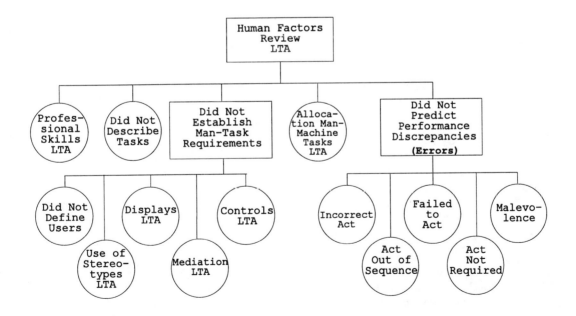

Figure 8–10. Human Factors Review LTA.

Did Not Establish Man–Machine Requirements
: *Does the review determine the special skills required from operators and machines?*

- Did Not Define Users
  : *Is available knowledge about potential users defined and incorporated in design?*
- Use of Stereotypes LTA
  : *Are checklists of stereotypes (typical, normal, expected behavior) used in design? (For example, is a control turned right to move a device to the right? Are controls coded by size, color, or shape?)*
- Displays LTA
  : *Are displays used which can be interpreted quickly with high reliability?*
- Mediation LTA
  : *Are delays and the reliability of interpretation/action cycles considered?*
- Controls LTA
  : *Are controls used which can be operated quickly with high reliability?*

Allocation Man–Machine Tasks LTA
: *Has it been determined at which tasks humans excel, and at which tasks machines excel? Has the determination been acted upon?*

Did Not Predict Performance Discrepancies (Errors)
: *Is there an attempt made to predict all the ways and frequencies with which human performance discrepancies may occur, and thereby determine corrective action to reduce the overall discrepancy rate?*

- Incorrect Act
  : *Have all the potential incorrect acts associated with a task been considered and appropriate changes made?*
- Act Out of Sequence
  : *Have the consequences of performing a task's steps in the wrong order been considered, and have appropriate corrective measures been taken?*
- Failed to Act
  : *Is there an attempt to reduce the likelihood of operators omitting steps or acts which are required by a procedure?*
- Act Not Required
  : *Are deliberate errors and other acts of malevolence anticipated and steps taken to prevent or reduce their effect?*
- Malevolence
  : *Are deliberate errors and other acts of malevolence anticipated and steps taken to prevent them or reduce their effect?*

Maintenance/Inspection LTA. The analytical logic for Maintenance and Inspection LTA in Design and Development is identical; it was presented in Chapter 7, The Strategies of Prevention by Safety Program Systems, Figure 7–10. Following is the added analytical comment that MORT presents on these two factors.

Is the maintainability/inspectability of an operation or facility given consideration during the conceptual (Design and Development) phase and on through the rest of the life cycle? Is there an adequate maintenance/inspection plan?

Arrangement LTA (Figure 8–8). *Does the design consider problems associated with space, proximity, crowding, convenience, order, freedom from interruption, enclosures, work flow, storage, etc.?*

Environment LTA (Figure 8–8). *Are people and objects free from physical stresses caused by (1) facility physical conditions; (2) conditions generated by the operation; or (3) interactions of one operation with another?*

Operational Specifications LTA (Figure 8-11). *Are there adequate operational specifications for all phases of the system operation?*

> Test and Qualification LTA
>> *Is there a "dry run" or demonstration to prove out all associated hardware and procedures and to check for oversights, adjust for the final arrangement, and provide for some hands–on participation?*
>
> Supervision LTA
>> *Are there guidelines for the amount of supervision required, the minimum supervisory capabilities needed, and the responsibilities of operating supervisors?*
>
> Personnel Selection LTA
>> *Are personnel selected on the basis of the capability (both physical and mental) which is necessary and sufficient to perform the operation?*
>
> Task Procedures Did Not Meet Criteria
>> *Do the procedures for each task meet selection and training criteria and applicable operating criteria? Are the procedures responsive to supervisory problems that can be addressed in written procedures?*
>>
>> Did Not Fit With Hardware Change
>>> *Are procedures revised, if necessary, to agree with changes in plant or equipment?*

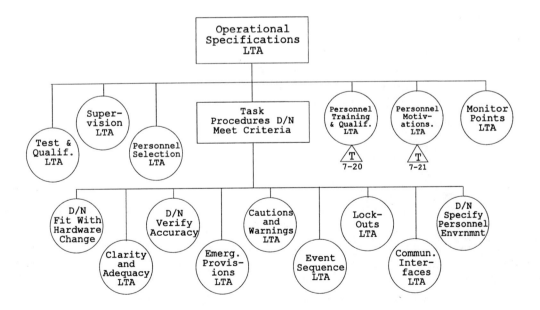

Figure 8–11. Operational Specifications LTA.

Clarity and Adequacy LTA
: *Are the procedures written to allow for the varying levels of reading skills and intelligence of the intended users? Do the procedures cover all steps of a task, and is enough information given about each step?*

Did Not Verify Accuracy
: *Are procedures checked and verified for accuracy under dry-run operating conditions?*

Emergency Provisions LTA
: *Do procedures give users clear instructions for all anticipated emergency conditions? Are instructions easy to perform under the stress of an emergency?*

Cautions and Warnings LTA
: *Are dynamic and static warnings used when appropriate? Are they located where the operation is performed as well as in procedures? Is their meaning unambiguous?*

Event Sequence LTA
: *Are procedures performed in a sequence that is according to criteria, is safe, and is sufficient?*

Lockouts LTA
: *Are lockouts called for where hazardous situations are encountered or created through the use of procedures?*

Communication Interfaces LTA
: *Do the procedures adequately convey their intended message?*

Did Not Specify Personnel Environment
: *Do procedures specify maximum permissible levels of physical stresses imposed on the users?*

Personnel Training and Qualification LTA
: *Are personnel given all the training they need for the equipment and procedures they will be using? Do they demonstrate through hands-on use that they know how to apply the training properly?*

  Personnel training and qualifications are considered in detail in Chapter 7, Figure 7–20.

Personnel Motivation LTA
: *Do personnel want to perform their assigned work task operations correctly?*

  Personnel motivation factors are considered in detail in Chapter 7, Figure 7–21.

Monitor Points LTA
: *Are there sufficient checkpoints in written procedures during an operation to assure that steps are being done correctly?*

Emergency Provisions LTA (Figure 8–9). *Does the design of plant and equipment provide for safe shutdown and the safety of persons and objects during all anticipated emergencies?*

General Design Process LTA (Figure 8–12). *Are commonly-recognized good engineering practices, including safety, reliability, and quality engineering practices, adequately incorporated into the general design process?*

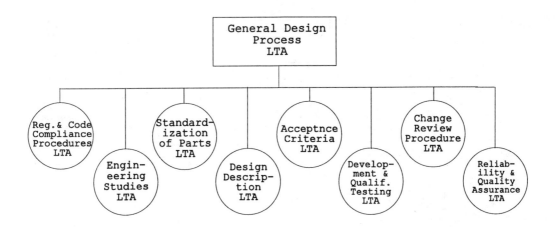

Figure 8–12. General Design Process LTA.

Code Compliance Procedures LTA
Are there written procedures to assure compliance with applicable engineering and design codes?

Engineering Studies LTA
Where codes, standards, regulations and state–of–the–art knowledge cannot furnish required design data, are engineering studies conducted to obtain the needed information?

Standardization of Parts LTA
Is there an attempt to use proven standardized parts where possible, or to design so as to encourage their use?

Design Description LTA
Does the design description provide all the information needed by its users in a clear and concise manner?

Acceptance Criteria LTA
Are acceptance criteria stringent enough to assure operability/maintainability and compliance with original designs?

Development and Qualification Testing LTA
Is there adequate testing during the development of a new design to demonstrate that it will serve its intended function? Does qualification testing assure that nonstandard components satisfy acceptance?

Change Review Process LTA
Does change review cover form, fit, and function on up the part–component–subsystem chain to a point where no change is demonstrated? Are there change dockets on drawings and at points–of–operation?

Reliability and Quality Assurance (RQA) LTA
Is there an adequate reliability and quality assurance program integrated into the general design process? (In some organizations, the RQA functions are very specifically separated; other organizations combine them. Whether combined or separated, RQA is a strong complement to safety. Close mutual support between safety and reliability and quality assurance should be evident throughout the general design process.)

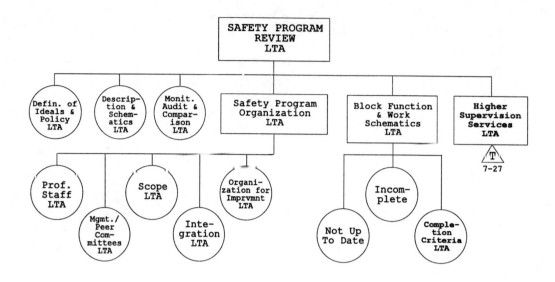

Figure 8–13. Safety Program Review LTA.

SAFETY PROGRAM REVIEW LTA (Figure 8–13)

Does the safety program review assure a planned and measured program, with low cost/high volume services, professional growth, and use of modern methods?

DEFINITION OF IDEALS AND POLICY LTA

Is there an adequate safety policy statement and are the ideals of the safety program articulated? Do these summarize what management should know (and require) of a safety program? Do the ideals provide a base from which to measure the program in order to project improvements?

The previous position on policy statements applies in this case as well. In quality managed organizations, separate functional policy statements are not needed.

DESCRIPTION AND SCHEMATICS LTA

Are program ideals documented in operating manuals and schematics? Are program operating data available and evaluated? Are there outlines, steps, and criteria which substantially describe the safety program?

MONITORING, AUDIT, AND COMPARISON LTA

Is there a formal measurement system which compares actual performance with safety program ideals and objectives?

SAFETY PROGRAM ORGANIZATION LTA

Is the program organized with the necessary and adequate elements?

Professional Staff LTA. Do safety personnel rate well by both safety and management criteria ? Are they effective in both technical and behavioral aspects? Do they have good organizational status and are they educated, experienced and promotable?

Management Peer Committees LTA. *Are special purpose and ongoing committees and boards used to improve safety understanding and attitudes within scientific, technical and engineering groups? Do these ongoing groups have a positive, action orientation toward real–life problems?*

Scope LTA. *Does the safety program address all forms of hazards, including anticipated hazards associated with advanced technological development and research?*

Integration LTA. *Is the staff support for safety integrated in one major unit rather than scattered in several places?*

Organization for Improvement LTA. *Is the safety program organized adequately to achieve the desired pace of safety improvement? (Achievement of a breakthrough goal in accident reduction by a safety program requires clear goal definition and distinctive organizational effort.)*

BLOCK FUNCTION AND WORK SCHEMATICS LTA

Are charts and drawings of the full array of safety–related processes and functions adequate and reviewable? (This may include providing safety equipment, delivering safety reviews to points of need, and other safety–related functions, plus the schematics of various "upstream processes" that are to be audited or monitored.)

Not Up-to-Date. *Are charts and drawings kept up–to–date?*

Incomplete. *Is everything that is needed for review included in the charts and drawings?*

Completion Criteria LTA. *Are criteria clear and specific as to what should be included in drawings and when they should be finished and revised?*

SAFETY PROGRAM SERVICES LTA

Does management provide the supportive services and guidance needed at lower organizational levels for an adequate safety program review?

Note the transfer in of all the lower tier events from Figure 7–27, Higher Supervision Management Services.

Oversights and Omissions vs. Assumed Risks

Chapters 7 and 8 have set forth a somewhat abbreviated version of the safety program systems and management safety systems established in the MORT as the content of the "What Happened?" and "Why?" questions of the major "Oversights and Omissions" branch of the risk tree.

However, in the top events of the MORT there is a second major branch, that of "Assumed Risks," as shown in Figure 8–14.

The text of the *MORT User's Manual* presents the following guideline questions for analysis of the assumed risk causation factor:

Assumed Risk:

What are the assumed risks? Are they specific, named events? Are they analyzed and, where possible, calculated (quantified)? Was there a specific decision to assume each risk? Was it made by a person who had the management–delegated authority to assume the risk?

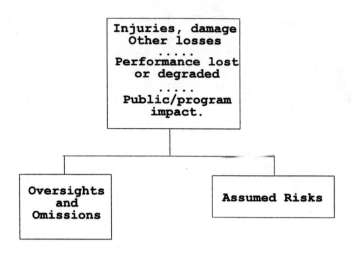

Figure 8–14. "Assumed Risks" is a second major branch of the Management Oversight Risk Tree, as shown.

The specific risk may be: (1) tolerably low (minor) in frequency or consequence; (2) high in consequence but impossible to eliminate (for instance, a hurricane); or (3) simply too expensive to correct when weighed against the risk consequences. The assumed risk is flagged with a small oval symbol lettered "AR".

Figure 8-15 presents the MORT summary for review.

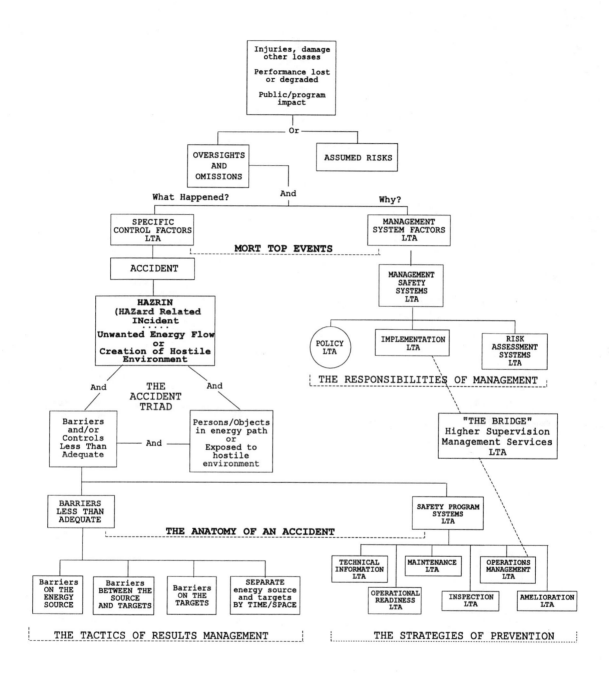

Figure 8-15. The Management Oversight Risk Tree Summary.

9

The Management Oversight Risk Tree Diagrams Restated

The introductory summary of the *User's Manual* for the Management Oversight Risk Tree is presented in the "Abstract":

> This report is the User's Manual for MORT, a logic diagram in the form of a "work sheet" that illustrates a long series of interrelated questions. MORT is a comprehensive analytical procedure that provides a disciplined method for determining the causes and contributing factors of major accidents. Alternatively, it serves as a tool to evaluate the quality of an existing safety system. While similar in many respects to fault tree analysis, MORT is more generalized and presents over 1500 specific elements of an ideal "universal" management program for optimizing occupational safety.[1]

The Management Oversight Risk Tree is a monumental document, but it seems apparent that the sheer size, complexity, and style have had a negative effect on its acceptance by the majority of safety professionals. "They have got to simplify it," is a common complaint. The presentation of "...over 1500 specific elements of an ideal 'universal' management program for optimizing occupational safety" has overwhelmed prospective customers.

As a practical matter probably no more that ten percent of the specific elements set forth in the MORT are vital to the development and operation of successful safety management programs in operations with risks of ordinary magnitude. Recognition of this reality has been one guiding principle in the development of the simplified restatement of the MORT presented herein. A second guiding principle has been rephrasing the elements into recognizable "safety systems," systems easily recognized as candidate subjects for continual improvement.

Chapters 7 and 8 displayed the definitions of events and cause factors that have been identified in the *MORT User's Manual* as potentially operative in accident or hazard related incident sequences. The cause and effect arrangement of the factors creates coherent and interdependent systems of incident causation. This identification of the systems of incident causation is the foundation upon which the objective of continual improvement of all safety systems, the practice of Total Quality Safety Management, rests.

In concept, the MORT is a disciplined method of analysis of unwanted energy transfers that produce, or have the potential to produce, wastes of all kinds: the waste of assets, the waste of materials, and the wastes of the time of people, all people, all levels. The MORT presents high–level ideals for safety programs and provides a format for program evaluation.

---

1. William G. Johnson, *MORT User's Manual* (Atlanta: EG&G Services, 1983).

This presentation introduces certain revisions of the original tree. Some are revisions of language that are designed to make the concepts more functional; to emphasize *what* was happening and avoid directing attention to *who* was involved. For example, "Supervision LTA" is renamed "Operations Management LTA".

Another example of language revision is the de–emphasis of the use of the word "error". In this revision the admittedly awkward term "performance discrepancy" is introduced. For many, the expression "to err is human" is a foundation concept, a concept basic to Taylorism and to the Heinrich theory of accident causation. The challenge here is not to introduce new and awkward terminology, but to depersonalize the understanding of the concept of error for everyone. In this regard Joseph Juran, in his book *Managerial Breakthrough*, presents a discussion of errors which is recommended reading.[2]

In addition to changes in terminology, major changes in the arrangement of the MORT logic are introduced. Every attempt has been made to more clearly identify coherent individual, yet interdependent, systems. Examples would be the systems of hazard detection and hazard correction; or the systems and sub–systems that are operative in the matter of performance discrepancies.

Closely related to these changes are changes in the MORT logic introduced by drawing a distinction between the elements of the safety program that are tactical in nature (directed to controlling the results of accident or hazard–related incidents) and those that are strategic in nature (directed to the prevention of unwanted incidents). The tactical elements are discussed in Chapter 6, "The Concepts of Energy and Barriers." The strategic elements of the safety culture are discussed in Chapter 7, "The Strategies of Safety Program Systems."

Chapter 8 is devoted to the "Responsibilities of Executive Management," clearly strategic in nature, constituting the basic nature of the entire organization, the "Why?" path of the analytical logic of the Management Oversight Risk Tree. At this point the operational connection between the "What Happened?" path and the "Why?" path is established in the system of Higher Supervision Management Services which serves as "The Bridge" between Management Safety System, (MSS), and the Safety Program Systems, (SPS).

The diagrams in the following pages are presented as system diagram work sheets which can be photocopied for use in diagnosis and analysis of hazard–related incidents, either realized or potential, and for evaluation of existing safety program practices and/or elements. An example of the use of the diagrams in the analysis of a case study accident is presented in Chapter 10. The elements in the diagrams are cross–indexed back to the respective diagram figure in the preceding text in Chapters 7 and 8. Figure 9–1 presents the MORT summary for reference.

Throughout it should be kept in mind that none of these presentations are final in form, but are presented as a pattern or formula that will enable continual improvement of the MORT itself. The diagrams are to serve as "working papers", used to guide discussions and deliberations of respective groups or teams seeking the identity and solution to safety system problems. In this regard comparison of this restated version of the *MORT Analytical Diagram* and *MORT User's Manual*,[3] or especially with the discussions presented by Johnson in his book *Safety Assurance Systems* is encouraged.

---

2. Joseph M. Juran, *Managerial Breakthrough, A New Concept of the Manager's Job* (New York: McGraw–Hill Book Co., 1964).

3. See footnotes 8 and 9, chapter 4.

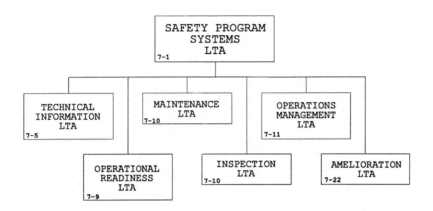

SAFETY PROGRAM SYSTEMS LTA
   TECHNICAL INFORMATION SYSTEMS LTA

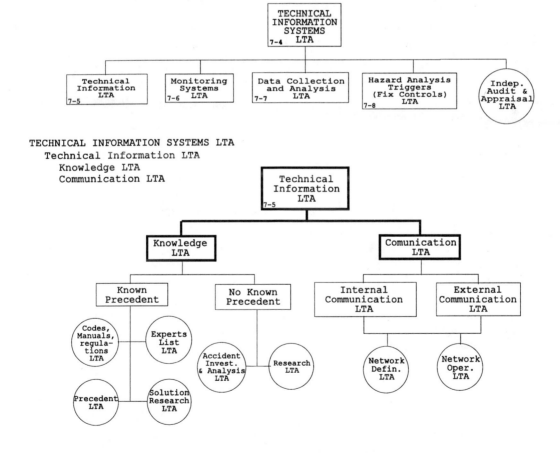

TECHNICAL INFORMATION SYSTEMS LTA
   Technical Information LTA
      Knowledge LTA
      Communication LTA

SPS–1. Safety Program Systems—the Strategies of Prevention.

TECHNICAL INFORMATION SYSTEMS LTA
    Monitoring Systems LTA
    Data Collection & Analysis LTA
    Hazard Analysis Triggers LTA

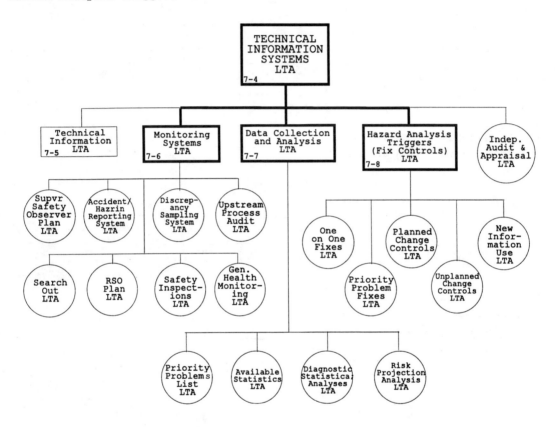

SPS–2. Safety Program Systems LTA: Technical Information Systems LTA.

OPERATIONAL READINESS LTA
    Verification of Use Readiness LTA
    Operations/Maintainence/Safety Interface LTA

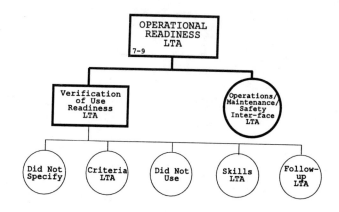

MAINTENANCE/INSPECTION LTA
    Plan LTA
    Execution LTA

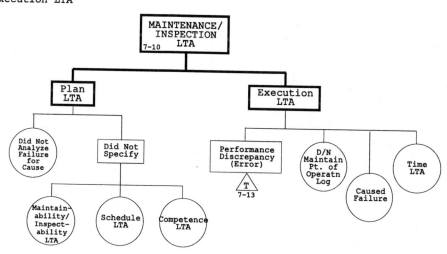

SPS–3. Safety Program Systems LTA: Operational Readiness and Maintenance/Inspection LTA.

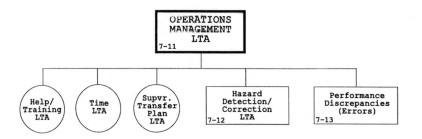

OPERATIONS MANAGEMENT LTA
   Hazard Detection/Correction LTA

      Did Not Detect
      Did Not Correct

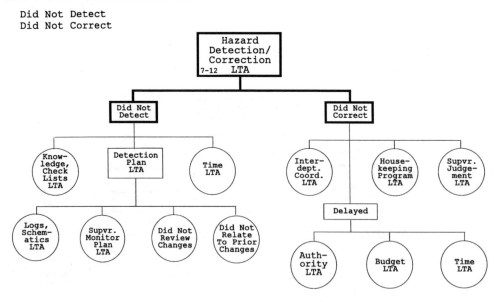

SPS–4. Safety Program Systems LTA: Operations Management LTA—Hazard Detection/Correction LTA.

PERFORMANCE DISCREPANCIES (ERRORS)

   Task Discrepancies
   Non-task Discrepancies
   Emergency Shut-off Discrepancies
     Task Performance Discrepancies

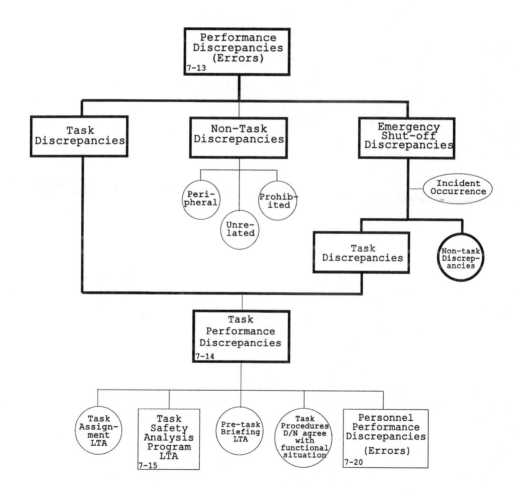

SPS–5. Safety Program Systems LTA: Operations Management LTA—Performance Discrepancies (errors).

Performance Discrepancies (Errors)
    Task Performance Discrepancies (Errors)
    Task Safety Analysis Program LTA
        Task Safety Analysis Not Performed
        Recommended Controls Not in Use

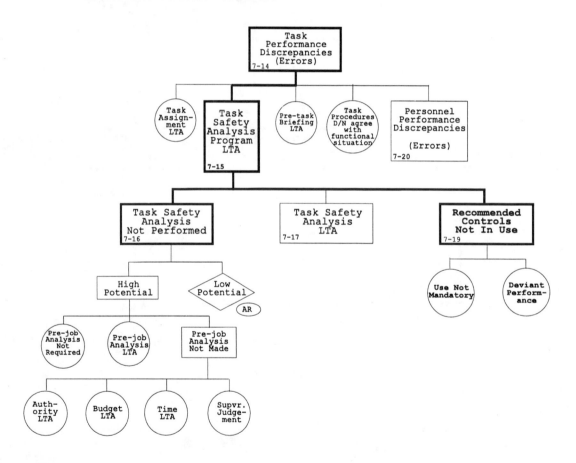

SPS–6A. Safety Program Systems LTA: Operations Management LTA—Task Performance Discrepancies (errors).

Performance Discrepancies (Errors)
  Task Performance Discrepancies (Errors)
    Task Safety Analysis Program LTA
      Task Safety Analysis LTA
        Preparation LTA
        Recommended Controls LTA

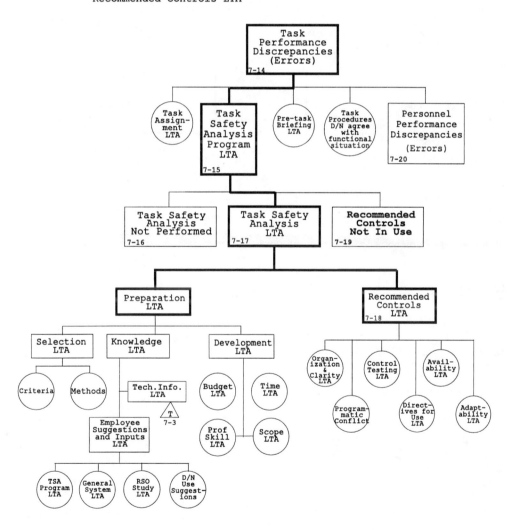

SPS–6B. Safety Program Systems LTA: Operations Management LTA—Task Performance Discrepancies (errors).

Performance Discrepancies (Errors)
   Task Performance Discrepancies (Errors)
     Personnel Performance Discrepancies (Errors)

       Personnel Selection LTA
       Training LTA
       Consideration of Deviations LTA

```
                    ┌─────────────────┐
                    │      Task       │
                    │  Performance    │
                    │ Discrepancies   │
                    │    (Errors)     │
                    │      7-14       │
                    └────────┬────────┘
      ┌──────────┬───────────┼──────────┬─────────────────┐
   (Task      ┌─────────┐ (Pre-task  (Task           ┌─────────────┐
   Assign-    │  Task   │  Briefing  Procedures      │  Personnel  │
   ment)      │ Safety  │    LTA)    D/N agree       │ Performance │
              │Analysis │            with            │Discrepancies│
              │Program  │            functional      │   (Errors)  │
              │  LTA    │            situation)      │    7-20     │
              │  7-15   │                            └──────┬──────┘
              └─────────┘                                   │
   ┌─────────────┬─────────────────┬───────────────────┬────┘
┌──────────┐ ┌─────────┐ ┌─────────────────┐ ┌─────────────┐
│Personnel │ │Training │ │ Consideration   │ │  Employee   │
│Selection │ │  LTA    │ │ of Deviations   │ │ Motivation  │
│   LTA    │ │         │ │      LTA        │ │    LTA      │
└────┬─────┘ └────┬────┘ └────────┬────────┘ │    7-21     │
     │            │               │           └─────────────┘
  ┌──┴──┐   ┌─────┼─────┐    ┌────┼──────┐
(Criteria)(Testing)(None)(Criteria)(Normal)(Did Not)(Did Not)
  LTA      LTA           LTA   Variability Observe  Correct
               (Methods)(Prof.              │         │
                 LTA    Skill            (Changes) ┌──┴──┐
                        LTA)                    (D/N   (D/N
                  (Verific-                    Rein-  Enforce)
                   ation                      struct)
                   LTA)
```

SPS–6C. Safety Program Systems LTA: Operations Management LTA—Task Performance Discrepancies (errors).

Performance Discrepancies (Errors)
  Task performance Discrepancies (Errors)
    Personnel Performance Discrepancies (Errors)

      Employee Motivation LTA

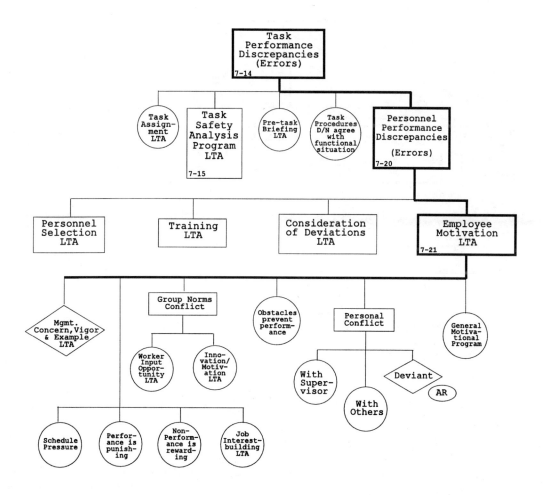

SPS–6D. Safety Program Systems LTA: Operations Management LTA—Task Performance Discrepancies (errors).

Prevention of Second Accident
Emergency Action LTA
Rescue LTA

Medical Services LTA
Rehabilitation LTA
Relations LTA

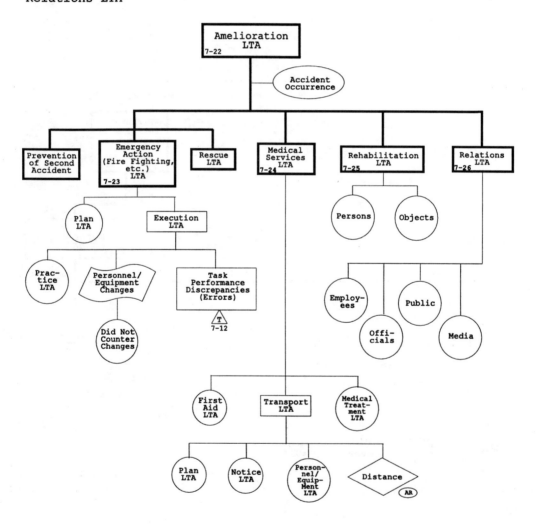

SPS–7. Safety Program Systems: Amelioration LTA (post–accident).

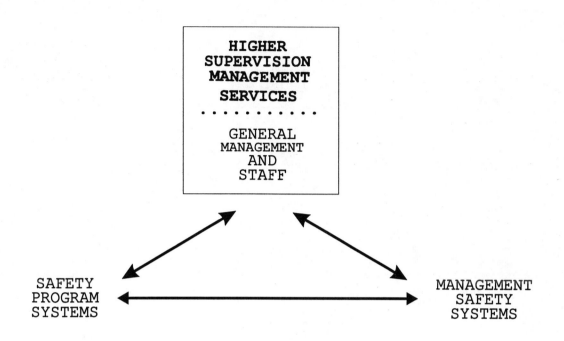

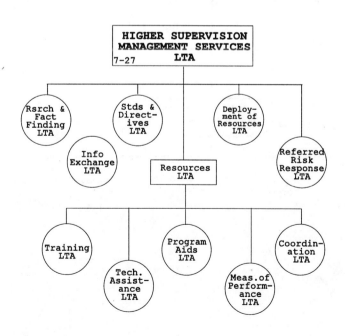

SPS–8. "The Bridge".

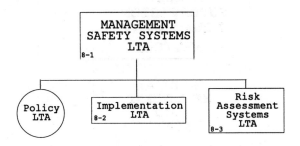

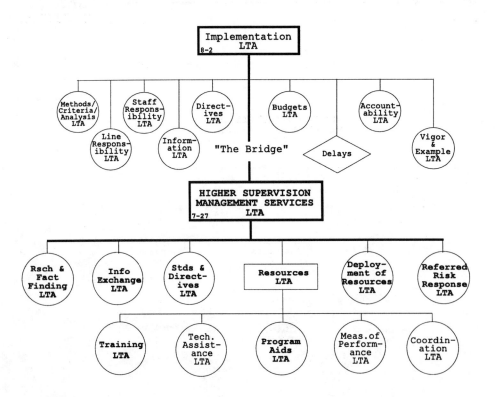

MSS–1. Management Safety Systems: Implementation LTA–Higher Supervision Management Services LTA.

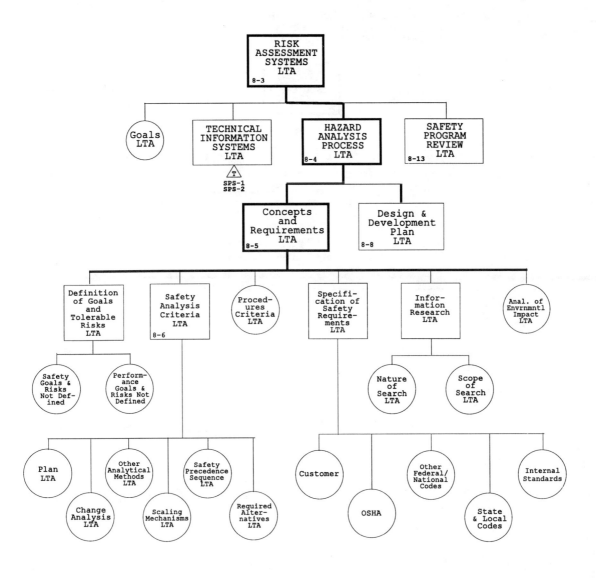

MSS–2. Management Safety Systems LTA: Risk Assessment Systems LTA.

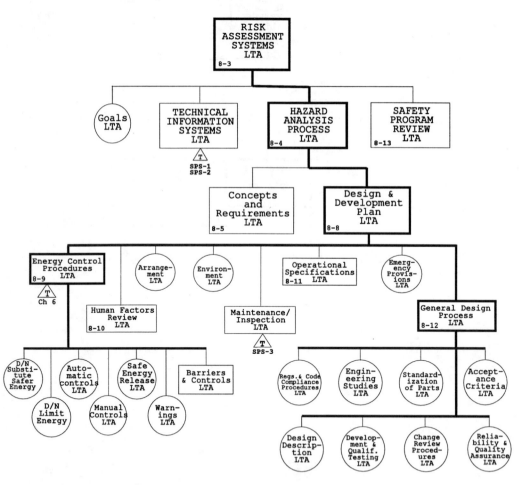

MSS–3. Management Safety Systems LTA: Risk Assessment Systems LTA.

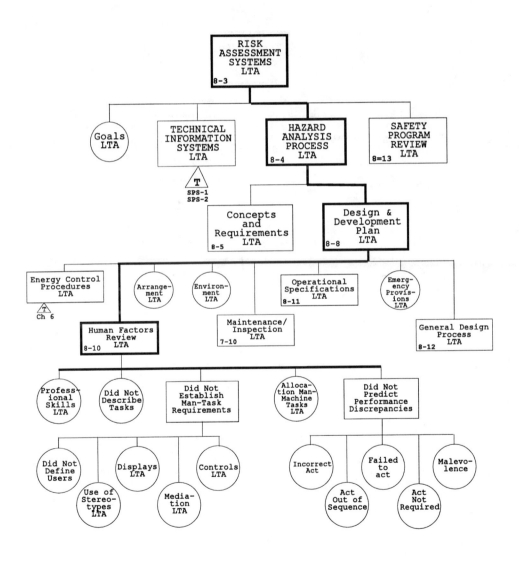

MSS–4. Management Safety Systems LTA: Risk Assessment Systems LTA.

MSS–5. Management Safety Systems LTA: Risk Assessment Systems LTA.

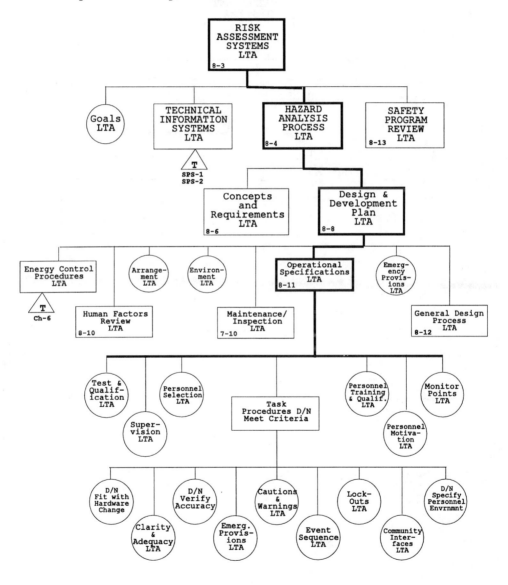

MSS–6. Management Safety Systems LTA: Risk Assessment Systems LTA.

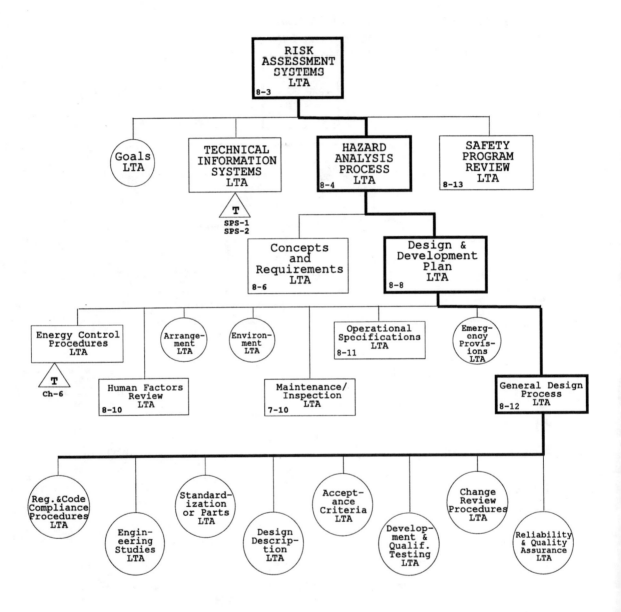

MSS–7. Management Safety Systems LTA: Risk Assessment Systems LTA.

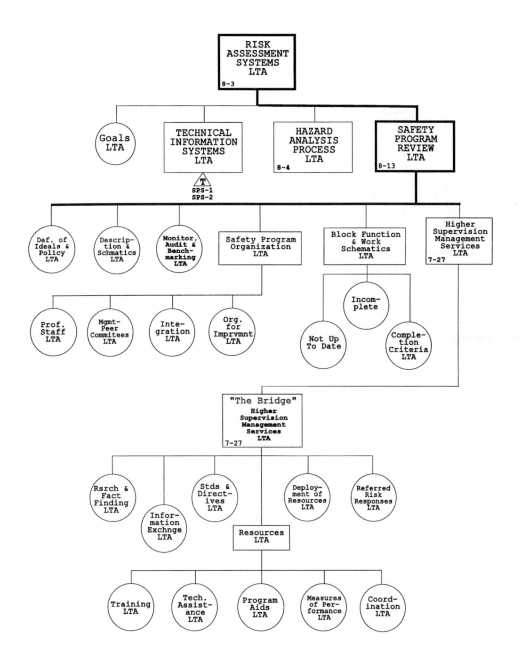

MSS–8. Management Safety Systems LTA: Risk Assessment Systems LTA and Safety Program Review LTA.

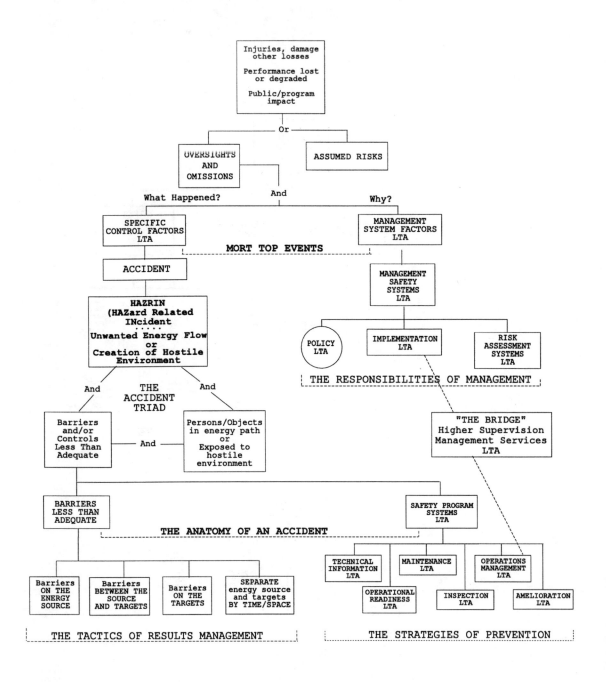

Figure 9–1. The Management Oversight Risk Tree restated.

10

Management by Fact
Achieving Continuous Improvement of Safety Systems by Statistical Methods

Dr. Deming's Challenge

At one of Dr. Deming's famous four day seminars, the question of the application of control charts to the problems of safety in the workplace was raised.[1] His response was quick and thunderous. "Forget about the charts. You must work on the causes!" He then expounded on the problems caused by improperly located directional signs on the nation's highways. His point however was clear.[2]

Dr. Deming's point was that figures on accidents do nothing to reduce the frequency of accidents. In order to reduce the frequency of accidents it is first necessary to determine whether the cause of the accident belongs to the system or to some fleeting cause, a specific person or set of conditions. Was the accident due to a common cause or was it the result of a special cause?

For many the first reaction is to attribute the accident to a visible something that someone did. This most often leads to the wrong answer, and more accidents continue to happen. The problem is confusing a special cause with a common cause. Action on the special cause leaves the common cause untouched, to return again for further trouble. Uncorrected common causes within the system guarantee that this will happen. Accidents due to common causes will continue to happen with an expected variable frequency until the system is corrected. Dr. Deming further pointed out that statistical methods provide the only method for developing an understanding of accidents that will enable their reduction.

The statistical method that will develop an understanding of the incident/accident process is the flow chart, specifically the Management Oversight Risk Tree, the fault tree flow chart that is the principal subject of this study. Generically the flow chart is a diagram that creates a picture of what happens in an ongoing system. As Dr. Deming makes clear, the chances of improving any system depends first on an understanding and agreement on what, in fact, has happened, is happening, or should be happening. For safety, this understanding is essential in determining whether the cause of the accident belongs to the system or to some specific person or fleeting set of conditions.

Flow diagrams can be created for any system or process. For instance, in her book *The Deming Management Method,* Mary Walton presents an amusing flow chart of "From Bed to Work".[3] The flow chart is a powerful diagnostic tool. It is not unusual to

---

1. The conference was held in St. Louis, Missouri, June 1991. The author was responsible for the question, which it later developed had been of interest to several others in the audience.
2. For a discussion on this point see: Dr. W. Edwards Deming, *Out of the Crisis* (Cambridge, Mass: Massachusetts Institute of Technology, Center for Advanced Engineering Study, 1986): 478–84.
3. Mary Walton, *The Deming Management Method* (New York: Dodd, Mead & Company, Inc., 1986): 103.

find that when different people, working in the same system, are asked to flow chart the way the system works, each constructs a different diagram. And when managers are asked to flow chart their operation they often find they must ask for help from their workers to learn what is really happening and who is responsible for what. Comparative flow charts of how a system should work and how, in practice, it does work can reveal misunderstandings and inefficiencies, oversights and omissions.

The Management Oversight Risk Tree is a specialized flow chart generally referred to as a *fault tree*. It starts with the result and works back to the causes of that result. The MORT is the key diagram for Total Quality Safety Management. The key action event in the MORT is the HAZRIN (HAZard Related INcident), the unwanted transfer of energy which includes the creation or existence of a hostile environment. As indicated in the Accident Triad (Figure 4–3) when the transfer occurs under conditions where the barriers and/or controls are less than adequate and people or objects are exposed to the energy transfer, wastes are the result. Expansion of the Barriers and Controls concept into separate categories of Barriers LTA and Safety Program Systems LTA revealed the Anatomy of an Accident (Figure 5–1). In this manner an accident is identified as a system or process that is ongoing when it occurs, a view that is consistent with the "systems vision" discussed in the preceding chapter. The past tense design of the MORT makes it particularly suited to the analysis of an accident or incident which has occurred.

Incident/Accident Investigations

The investigation of injuries has always been a priority concern of safety programs. Generally, the responsibility for making the investigation and report has been assigned to immediate supervisors as an integral part of their "key man" role in the safety program. This role assignment is an extension of the propositions that 80% or more of injuries are the result of the unsafe actions of the injured person or others, and that it is the supervisor's role to control the actions of the people.

The new paradigm of accident causation directly contradicts these propositions. Quality management recognizes that 85% or more of the variations that occur in ongoing systems arise out of common causes that are built into the system; that variation is normal in all systems or processes; and that controlling these variations can be effected only by management action to improve the system.

Quality management also recognizes that the remaining 15% or less of the variations arise from special causes from outside the system. The actions of the people working in the system are guided by the multiple systems in which they work. Behavior that might at first appear to fall outside of a particular system can only be classified as a special cause after full investigation of all systems that are operative in regard to that behavior have been considered and found *not* to have been a factor. This is the only circumstance under which the expression *unsafe act*, as most commonly understood, has validity. And, under quality management it is the manager's job to work on the systems within which the people work, to improve them, with the help of the people.

In the investigation of an accident or significant hazard related incident, the identification of the causes of undesirable variations must be held rigidly in mind. This is precisely the point that Dr. Deming was making. The occurrence of the HAZRIN has identified the variation. One value of control charts rests in the identification of variations in ongoing systems or processes that might otherwise go undetected by other means until it was too late. This is not a problem in the accident situation. Control charts would do nothing to help identify the causes of highway accidents. In inci-

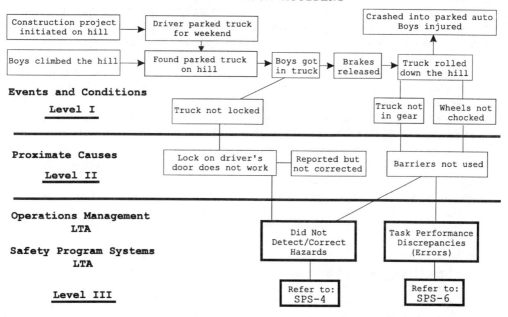

Figure 10–1. Runaway truck accident flow chart.

dent/accident investigations and analyses for cause the flow chart is the primary statistical method, the place to begin.

The Runaway Truck Accident

These principles, and the power of the flow chart as an analytical tool, can be demonstrated in the analysis of a relatively simple, yet serious, accident involving a runaway truck.

In this accident, a construction project had been started on a hill. The equipment left on the site for the weekend included a gravel truck. On Sunday afternoon two boys from the neighborhood climbed the hill, found the truck and got in to play truck driver. Somehow the brakes were released and the truck rolled down the hill. At the foot of the hill it struck a parked car, which was extensively damaged. The boys were both injured, one seriously. The neighborhood reaction to the incident was highly critical of the construction company.

Investigation of the events and conditions of this accident revealed that the driver's door of the truck had not been locked. In addition it was established the truck was not in gear, nor were the wheels chocked to prevent movement, although wheel chocks were carried on the truck. Further investigation of the proximate causes of the incident revealed that the lock on the driver's door of the truck was not working. It developed that this fact had been reported to maintenance but no action had been taken. And it was clear that the requirement to leave the unattended vehicle in gear and chock the wheels under these circumstances had not been followed.

These events and the immediate and proximate causes surrounding them are shown in the top two levels of the flow chart in Figure 10–1. Level I depicts the sequence of events and the conditions that allowed these events to occur. Level II displays the proximate causes, the causes that established those immediate conditions.

At this point no analysis for system causes of the accident have been made. Corrective action taken on the basis of the facts on hand at this point can do nothing to improve the system causes, the oversights and omissions involved. They have not been identified. That identification will occur only as the result of system analysis—Level III of the process.

In the conduct of the analysis it is intended that reproductions of the pertinent Safety Program Systems (SPS) diagrams will be used to track the logic. The process is one of identifying basic problems, those elements identified in the circles or bubbles. Johnson[4] suggests the use of colored pencils: red for the bubbles that appear deficient; green for those that are okay; and blue for bubbles where the answer is not clear and further study is needed for clarification.

In the early stages of the analysis the effort will concentrate on the specific factors of Safety Program System inadequacies. As the analysis develops understanding of relevant SPS inadequacies, questions about Management Safety System (MSS) factors that may be inadequate may develop and require consideration. These questions will most often arise in the generic Operations Management LTA safety program system (SPS–12) which has been identified as the "bridge" between Safety Program Systems and Management Safety Systems.

Johnson suggests that the normal expectation is to get some reds (obvious inadequacies) and a great many blues. In using MORT for accident analysis not many greens will appear, because at this time the procedure is not for the purpose of an overall safety program appraisal. The process is complete when the bubbles in all SPS diagrams are colored either red or blue or green, or the SPS diagram has been set aside as not relevant to the situation.

In the runaway truck example the relevant SPS diagrams were identified as SPS–4, Operations Management LTA: Did Not Detect/Correct Hazards, and SPS–6, Operations Management LTA: Task Performance Discrepancies. Once a bubble has been colored red, that red assignment flows up through the system–generic factors of which the specific factor is a part. System improvement will occur as the blues are converted either to red or green through study, and action is taken to convert the reds to green.

In practice it may develop that the circumstances of the incident being analyzed do not fit well into the MORT pattern of factors. If so, constructing a tailor–made diagram should be considered, flow charting the incident and converting it to a logic tree. The MORT, as presented, is a method, not an answer. It is open–ended in principle. It may also occur that something that the MORT indicates is a deficiency may not be considered a deficiency in the culture of some organizations. It is not intended that deficiencies be called out simply because the MORT classified it as LTA. However, the validity and value of a contrary finding or recommendation must be defendable. The objective of the MORT is to indicate good places to look for deficiencies, to bare the potentials for oversights and omissions.

---

4. William G. Johnson, *MORT Safety Assurance Systems* (New York: Marcel Dekker, Inc., 1980). Johnson presents the most complete discussion and directions on the uses of the MORT charts. The student is urged to consult this source for comprehensive coverage of the subject.

The Need for Scaling Devices (the Pareto Principle)

The Management Oversight Risk Tree is an intense and complex analytical procedure. Is intensive analysis of *all* accidents the expectation? The answer is that it is not.

Incidents and accidents vary widely in their complexity and potentials for harm. The greatest value will be derived from the MORT discipline through the intensive analysis of a few serious incidents. It is in the serious incidents that the opportunities to learn and opportunities for system improvement are the greatest.

The problem that arises is quantifying the seriousness of incidents so that these opportunities are not missed. In situations where the injuries are serious or the damages are extensive, there should be no question about the need for in–depth investigation and analysis. But improvement and learning opportunities are not restricted to such obvious situations. Not infrequently, there is as much or more to learn from incidents in which the third element of the triad was not operative and little or no harm to persons or objects occurred, the familiar, but poorly titled, "near–miss".

In the literature of safety little attention has been paid to the matter of scaling the depth of investigations to the seriousness of the incident, scaling the potential for harm, or the potential for learning and improvement from the occurrence. Traditional safety wisdom has insisted that all injuries be thoroughly investigated since the severity of any injury is largely a matter of luck. The determination that the key element in the accident is the unwanted transfer of energy renders that theory of injury severity moot; it is not luck that determines severity, it is the amount of energy that reaches the target, whether the targets are people, objects, or the ongoing operation.

The need for scaling devices is mentioned in the MORT within the Management Safety Systems diagrams. Diagram MSS–2 is devoted to the generic Management Safety System of Risk Assessment. In that diagram one major subsystem is the Hazard Analysis Process (HAP). That analysis is divided into two paths, Concepts and Requirements of the HAP, and the Design and Development of the HAP. In the systems of Concepts and Requirements the adequacy of Safety Analysis Criteria is reviewed. It identifies Scaling Mechanisms LTA as a Specific Factor failure. Reference to the analytical logic of Safety Analysis Criteria LTA, Figure 8–6, reveals the following guiding questions for Scaling Mechanisms LTA:

- Has some reasonably clear–cut mechanism been established for scaling the seriousness and/or the severity of prior events?
- Is there a mechanism to project past events to a scaled effort to evaluate current practices?

In spite of the urging of some that every injury is critically important, experience shows that this simply is not true. The ability to recognize those that are significant will assure concentration of effort on significant problems and avoid the promotion of trivia into problems or creating hobby type projects, or worse yet, using the activity on trivia as cover for failure to determine and act upon the true cause.

The statistical principle that underlies the concept of scaling is widely recognized as the Pareto Principle or as the Principle of the Vital Few. Credit for the application of the Pareto Principle to quality management is given to Joseph M. Juran.[5] It was he who demonstrated and expanded on its statistical validity to render it one of the most important principles of management by fact. A specific reference to the application of

---

5. Joseph M. Juran, *Managerial Breakthrough, A New Concept of the Manager's Job* (New York: McGraw–Hill Book company, 1964): chapter 4.

this principle to accident/incident investigations is made by Kuhlman in his book, *Professional Accident Investigation:*[6]

> "In any group or array, a relatively small number of items will tend to give rise to the largest proportions of results.
>
> "As startling as it might seem to some, this principle indicates that only one problem out of four is worth more than a loss control manager's fleeting glance. The other three can be dealt with summarily, since they are rarely of significant consequence. The loss control manager who recognizes this well–documented but often overlooked economic (statistical) law takes a giant step in directing his limited time and efforts where they will do the most good... .
>
> "This principle...is so wide and its value so great in terms of time and effort conservation that it is considered one of the most important management principles."

The Classified Accident/Injury Investigation

There is one ability that every successful safety practitioner develops through experience that contributes substantially to that success. This is the ability to quickly form a judgment on the implications of any incident/accident situation. Almost instinctively an opinion will be formed as to whether or not a given incident is one with significance for the prevention effort. "Is this ability, gained through experience, a teachable skill?" is a proper question that arises.

An example of the development of a scaling device is one that centered on rating injury accidents in a high frequency/low severity exposure.[7] The project was initiated with a question: "What are the specific factors about any given injury accident that create the almost instinctive reaction of significance or lack of significance?" In the deliberations five factors were selected as being most influential. They were:

1. How bad was the injury?
2. Who was injured?
3. What level of energy was involved?
4. What type of equipment was involved?
5. What type of material was involved?

The next step was to create five categories of increasing intensity or severity under each of the five factors. Each category was given a power rating ranging from 0–10 or 0–15, and the power rating ranges were given titles. The titles in ascending order were *Nonserious, Serious, Severe,* and *Critical.* A fatality or total permanent disability was automatically rated at 50 points and categorized as a *Catastrophe.*

The remaining task was to describe each category for each factor with language that conveyed to the rater an understanding or feeling that would help in assigning an intensity or severity point value. The resulting Guide to Accident/Injury Rating is shown in Figure 10–2.

---

6. Raymond L. Kuhlman, *Professional Accident Investigation* (Loganville, GA: Institute Press, 1977).

7. Edward E. Adams, "Accident Investigation Procedure—Some Guidelines for Classification," *Professional Safety* (August 1985).

GUIDE TO ACCIDENT/INJURY RATING

	CRITICAL 9-10 POINTS	SEVERE 6-8 POINTS	SERIOUS 2-5 POINTS	NON-SERIOUS 0-1 POINTS	PTS.
INJURY SEVERITY	Major member loss (foot, hand, arm eye) Long term disability (More than 30 days)	Loss of lesser member or impairment of function Extended lost time (more than 7 days)	Extensive medical with or without lost time Objective evidence of pain or injury	No visible injury No clear incident Minor medical (First aid or one MD trtmnt)	___
PERSONNEL INJURY	Manager or Supervisor	Long-term technical or skilled employee New or lead employee	New employee Recently transferred employee	All other employees	___
INJURY POTENTIAL BY ENERGY LEVEL	Contact with a level of energy that threatens survivability of the body or its part	Contact with a level of energy well beyond the threshold limits of the body	Contact w/moderate level of energy, but beyond threshold limits of the body	Contact with very low level of energy No evidence of contact with energy	___
EQUIPMENT TYPE	Production machines, mobile powered equipment, or other powered, energized or presurized equipment Power transmission Point of operation Electrical apparatus	Moving equipment Elevating or conveying equipment Elevated scaffolds or work platforms	Hand tools Non-powered equipment Portable electric or air operated equipment Ladders or stairs	No equipment involved	___
MATERIAL TYPE	Highly corrosive Lethally toxic	High thermal Mildly corrosive Mildly toxic No Materials involved (6 points)	Sharp-rough-pointed-slippery, incl. floors Heavy: Males— over 50 lbs. Females— over 30 lbs.	Non-hazardous	___

Fatality or Permanent
Total Disability: 50 Points

Accident/Injury
Rating: **CATASTROPHE**

TOTAL POINTS: _____

ACCIDENT/INJURY RATING: _____

Figure 10-2. Guide to Accident/Injury Rating.

Some comments on the descriptions of the individual factors can be made:

Injury Severity: How bad was the injury? This proved to be the easiest factor to evaluate. The language seemed to fit well and developing a "feel" for the answer was not difficult for most.

Key Personnel Injury: Who was injured? Here the answer turned on the questions of role model, experience, and training. The highest rating, *Critical,* is assigned to the manager or supervisor who are injured. The second highest rating, *Severe,* is again a matter of being a role model. The phrase *key employee* is interpreted as an employee that other employees consciously or unconsciously look to: for example the employee selected to train new employees. Like supervisors, these are people who should not get themselves injured.

The *Serious* category relates to training and experience. And, finally there are all the other employees, the most populous group. The low rating does not indicate lack

POINTS	A/I RATING	INVESTIGATION BY	INVESTIGATION PURPOSE AND TYPE OF REPORT
50	CATASTROPHE	Blue Ribbon Team V.P. Production	Level III Investigation Basic Cause Determination Identification of oversights and omissions in Safety Program and Management Safety Systems Employing statistical methods of analysis; develop recommendations that will improve safety systems identified as less than adequate. Assign responsibility for implementation and project dates for accomplishment in the narrative report to executive management.
40-49	CRITICAL	Location manager Department manager(s) Immediate supervision Task-skilled employee(s) Technical experts Safety management	
26-39	SEVERE	Red Ribbon Team Location Manager Department manager(s) Immediate supervision Task-skilled employee(s) Safety supervision	
10-25	SERIOUS	Department Team Department Manager Immediate supervisor Task skilled employees Safety supervisor	Level II Investigation Investigate and correct immediate and proximate causes. Safety to monitor frequency of incidents by cause for evidence of safety program systems inadequacies.
0-9	NON-SERIOUS	Medical Department Safety Supervisor	Record of incident or complaint and of treatment rendered. Monitor frequency of incidents or complaints for evidence of safety program system inadequacies.

Figure 10-3. A/I Rated Investigation Procedures.

of their importance, but rather a nonsignificant influence in the matter under consideration.

Injury Potential by Energy Level: What level of energy was involved? This proved to be the most difficult factor to describe. It is also the factor with the highest sense of redundancy. Yet it is the most essential of all five factors. The discussion of Energy in Chapter Six is helpful in evaluating this factor.

Equipment Type: What type of equipment was involved? This factor was described by using examples that were scaled upwards with increasing energy levels. Equipment Type should be interpreted more broadly than Agency. Collapse of a tiered stack in the warehouse could involve forklift equipment even though the truck itself had departed the scene.

Material Type: What type of material was involved? The most significant and most surprising finding here was the importance attached to "No materials involved" in practically every such case. This was determined in trial ratings of over two hundred accident reports, and later reconfirmed in actual practice. An arbitrary six points was assigned to this finding.

This statistical method of separating the vital few from the trivial many enables classification of the accident by total points, so that the investigative procedure can be scaled to the significance of the incident. The scaling mechanism for the investigation procedure is presented in Figure 10-3.

The Accident/Injury Rated Investigation Procedures are designed to establish the scope and intensity of the investigation of those accidents which result in injury to persons. As presented in the MORT "Scaling Devices LTA" is phrased in the plural, indicating the probability that different versions may be necessary for different types of incidents. Adjustment of the categories to suit the requirements of no–injury accidents, accidents that result in damage to assets and/or degradation of the operating system should not prove difficult. The major adjustment required would appear to be in the Injury Severity category, replacing it with expressions that would rank Damage Severity. However, the suggestion is that significance be assigned to the "Potential for Injury to Persons" even though no injury had occurred.

Application of the statistical method of the Pareto Principle to the investigation and analysis of both accidents, where the triad had progressed to completion, and incidents, where only the first element of the triad, the unwanted energy transfer, had occurred, is essential to the Total Quality Safety Management program. This is the significance of the MORT subsystem of "Scaling Devices" as a factor in the Management Safety System of the Hazard Analysis Process.

Team Investigations

When an accident of catastrophic proportions occurs, multi–disciplined team investigation and analysis has always been considered essential. The greater the magnitude of the event the greater the team, the greater the number of discipline experts, the more intense the attention given to even the smallest detail, and the more complete and detailed the final report. But, just as the enormity of the event increases the enormity of the investigation/analysis effort, decreasing enormity of the event quite naturally results in decreased effort. This decrease goes down in scale to the point where many traditional safety programs assign the effort to one lone individual, the first line supervisor, whose responsibility it is to complete "The Supervisor's Injury Investigation". The difficulty is that too often this assignment is beyond the scope of the supervisor's authority to influence the situation (it is interdepartmental in nature), or beyond his knowledge level for effective analysis, or beyond the time demands of matters he believes more important in the eyes of his superiors. And, there is a final difficulty that must be recognized. That difficulty arises when even cursory investigation reveals the strong likelihood of mismanagement, of management or supervisory discrepancies (oversights or omissions) that could prove embarrassing. (No one ever works very hard at proving himself or his boss wrong.)

The principle of scaling the investigation procedure to the seriousness of the event is the basis of the Accident/Injury Rated Investigation Procedures chart (Figure 10–3). Their construction introduces the need for broadening the scope of the procedures in order to increase inputs from expanded levels of involvement. This expansion goes in both directions, upward to higher management levels, and downward to the level of the people. These expansions redefine the investigative effort into a team project, a team responsibility.

The Team Responsibilities of Management

In the restatement of the Management Oversight Risk Tree diagram, SPS–11 characterizes Figure 7–27, Higher Supervision Management Services, as "The Bridge" between

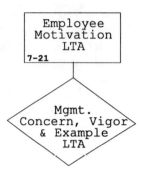

Figure 10–4. Diagram of the most critical management "service" of all, in SPS–9.

Safety Program Systems and Management Safety Systems. Examination of the analytical details of Figure 7–27 reveals that all of the factors listed are, in a sense, technical services. As a result, one management "service" is not included in the display, and for the management of safety programs it is the most critical management service of all. That service is visible demonstration of support at all management levels for safety and the continual improvement of safety program and management safety systems.

This critical issue is encountered however, in the link traced in SPS–9, from Operations Management, to Performance Discrepancies to Employee Motivation LTA. One of the principal factors shown under this subsystem is diagrammed with the diamond–enclosed legend shown in Figure 10–4.

The MORT legend for the diamond figure is given as:

> "An event whose sequence is terminated for lack of information, consequences, or solution. The event may be transferred to Assumed Risk."

The MORT legend for Assumed Risk is given as:

> "Transfer to Assumed Risk for problems for which there is no known countermeasure or no practical countermeasure."

In MORT practice, assignment to Assumed Risk is restricted to risks that have been identified, discussed, and evaluated, and a deliberate decision has been made to undertake the risks involved. The neglect of effective management demonstrations of concern, vigor and example in regard to the conduct of the safety program may not be a risk that has been consciously assumed. If this is so, the event must then be classed as a management oversight or omission. There are practical countermeasures available, and one powerful measures would be active, visible, participation in the accident/incident investigation processes. Management "policy statements" relative to the importance of safety have no real effect in this matter. Policy is not what is said, it is what is done. Application of the Pareto Principle in this matter is the suggestion.

The Accident/Injury Rated Investigation Procedures

In the A/I Rated Investigation Procedures (Figure 10–3), scaled investigation procedures are presented. At the very lowest level of incident severity, rated Nonserious, a logged record of the event that includes information about the incident, made at the time of treatment, is judged sufficient. The issue at this level is not severity, it is frequency. Within the total range of incident severity, the Nonserious category falls into the Pareto classification of the "Trivial Many." Monitoring of frequency by the medical department, assisted by safety supervision, is adequate.

At the other four levels of incident severity, the opportunity for visible and meaningful demonstration of management concern, vigor and example presents itself. In the "Investigation By" section of the procedures rating chart, three different investigating teams are suggested—the Departmental Team for the Serious rating, a Red Ribbon Team for the Severe rating and a Blue Ribbon Team for the categories rated Critical and Catastrophe. Within each of the team categories, the team leader is listed above the suggested team membership. As the intensity of the investigation rises so does the management rank of the individual assigned responsibility for the investigation.

The Departmental Team (Serious Rated)

The Serious rated incidents constitute the remaining portion of the Trivial Many. The reasons for this classification will become clear as the discussion moves into the higher rated categories. The Nonserious and Serious categories in combination account for about 75% of the total incidents that occur in any statistically significant period of time.

Responsibility for the investigation at the Serious level is assigned to the department manager, the individual who has a direct reporting relationship to the plant or location manager. The suggested team members include the supervisor of the injured, one or two task skilled employees, and the individual assigned responsibility for the location safety program, either full-time or part-time. Others may be asked to participate as is felt necessary, for example, maintenance supervision.

Again, at this level of severity, frequency of occurrence will be more the problem than injury severity. And it is not unusual to find that most of the problems at this level of severity center on the safety program system of barriers being less than adequate. However, it is always important that the team be alert to indications of problems that go beyond such proximate causes, for example, problems of hazard detection and/or correction, or problems that arise out of inadequacies in task safety analysis and training.

In the management of Accident/Injuries rated Serious it is the responsibility of location managers to hold their immediate staff responsible for compliance with both the letter and the spirit of the program. Location managers are responsible for projecting management concern, vigor and example.

THE RED RIBBON TEAM (SEVERE RATED)

Responsibility for the investigation at the Severe level is assigned directly to the location manager. At this point the concern starts to shift into the range of the Vital Few. Of the total number of accidents investigated over a significant period, only about 25% will be classified Severe or greater to qualify as belonging to this significant grouping. In approaching the problems of the Vital Few, Dr. Juran has offered pertinent comments on their characteristics.[8] He points out that while each of the Vital Few is an important and unique problem, they possess some common characteristics. These characteristics bear significantly on the matter of accident incidents and the investigation/analysis system.

---

8. Juran, *Managerial Breakthrough, A New Concept of the Manager's Job*: 52.

His first comment is that problems classified as vital are, almost without exception, not only unique but interdepartmental in character. And, he continues, since each is a unique problem, it requires study in depth if a solution is to be found. That study must involve much "digging in the field". While it may be possible to identify a vital problem from afar, the solution requires the details and the knowledge about events and causes that is only available at the action scene, from the only people who have that firsthand knowledge essential to success. Dr. Juran concludes that two needs must be met to achieve solution of the Vital Few: the implications of interdepartmental involvement must be met; and problem solution based on in-depth investigation and analysis is required. Identification of the problem is not adequate

Dr. Juran also warns that it is exaggeration to state that the Pareto Principle sorts things into two neat piles. There are, in reality, three piles; the third, or middle, pile being the residue that falls between the Vital Few and the Trivial Many. It is into this category that the Severe rated incidents fall. He labels this the "awkward zone". The individual incidents are not big enough or unique enough to justify tailor-made analysis for each, yet they are not collectively small enough to be treated as one homogeneous mass. As competence is gained in the analysis and treatment of the truly Vital Few and in the analysis and treatment of the Trivial Many, the result will be increased ability to handle "the awkward zone"

The composition of the Red Ribbon Team seeks to reflect the concerns expressed by Dr. Juran. In the first instance, responsibility for the investigation and analysis at the Severe level is assigned to the highest management level on site, the location manager. This is essential for the projection of management concern, vigor and example. It is also essential due to the very strong probability that the problem will be interdepartmental in character. As the investigation/analysis progresses, following the MORT analytical diagrams, and interdepartmental relationships are identified, the participation of the managers of those departments becomes necessary. System change is the anticipated solution, and this will involve managers at this level. Their participation in the deliberations will be essential for successful development and implementation of the changes.

The participation of immediate supervision and of task skilled employees is required for the "digging in the field", obtaining the essential first-hand knowledge about events and causes that is known only to those working in the system or process. The keenest observers of interdepartmental conflict, of conflicting demands made by different systems, are the workers, and (perhaps to a lesser extent) the immediate supervisor. These knowledge sources of "how things really are" will be the source of the causes, the details, and the discovery of the new knowledge required for effective solution.

Participation by the location safety supervisor (hopefully a trained safety resource) would serve as a technical safety expert and should be expected to serve as secretary for the proceedings.

THE BLUE RIBBON TEAM (CRITICAL AND CATASTROPHE RATED)

The essential difference between the Red Ribbon Team and the Blue Ribbon Team is assigning responsibility to a higher level of management. The team leader listed is the Vice President of Production. The interpretation should be "the individual in general management to whom the location manager reports." In other words, at this level of severity, general management becomes responsible for the conduct of the investigation/analysis and eventual solution. At the Critical level, de facto direction may be delegated to the location manager, with close advice to the manager's superior. However, the team members should know that the leader is, in fact, the executive and they are responsible to him for results. At the Catastrophe level it is expected that the executive would personally lead in the establishment of the Blue Ribbon Team, initiate the discussions, maintain close contact on progress and personally participate in the solution deliberations. This level of participation would be a minimum demonstration of management concern, vigor and example.

The makeup of the Blue Ribbon Team follows the pattern established for the Red Ribbon Team, except that increased assistance from technical experts, internal or external, broader interdepartmental participation, and assistance from the highest level of safety management would be anticipated.

At both the Red and Blue level, the analytical logic of the Management Oversight Risk Tree, or some equally specific and detailed guide to analysis, must be employed to assure management by fact. Unstructured, nonspecific discussion and deliberation will have little guarantee of success, could result in misguided decisions, and most certainly will consume excessive amounts of time.

The Pareto Distribution

A summary diagram of the Vital Few and the Trivial Many is shown in Figure 10–5. Adapted from a presentation of the Pareto Distribution shown by Dr. Juran,[9] the chart summarizes the points made in the application of the Pareto Principle to injury investigations.

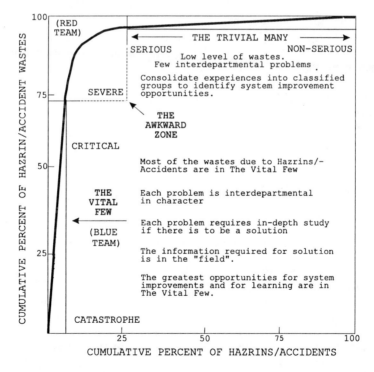

Figure 10–5. The Pareto Distribution.

---

9. Juran, *Managerial Breakthrough, A New Concept of the Manager's Job*: 53.

The Pareto Redistribution (from Trivial to Vital)

The Pareto Distribution and the discussion from which it evolved is founded on non-mathematical data. The values established in the Guide to Accident/Injury Rating are subjective in nature, the result of thought processes based on experience, observation, and deliberation, and tested against recorded experience for reasonableness. A proper caveat would be that the guide presented here is not a formula for universal application, but rather it is a suggested pattern for the development of tailor-made scaling devices for other types of exposures.

The A/I Rating Chart enables the separation of injury into three separate piles:

1. The Vital Few (low frequency–high severity), where about 10% of the occurrences result in about 75% of the total wastes of accidents.
2. The Trivial Many (high frequency–low severity), where about 75% of the occurrences result in about 10% of accident wastes.
3. The awkward zone (moderate frequency–moderate severity), where about 15% of the occurrences account for the remaining 15% of the wastes.

In discussing the Trivial Many, Dr. Juran points out that each of the trivial is by definition inconsequential, and since it is not feasible to deal with them individually, it is necessary to group them into classes. Since the problems are those of frequency rather than severity, classification results in data that is numerical, objective data developed by simple counting that can be plotted as a chart. The resulting statistical diagram is one of the most commonly used graphic techniques, the Pareto Chart.

Classifying the data into groups for comparison of the frequency of occurrence establishes priority ranking to guide corrective action efforts. However, it does more than establish priority. It can result in transforming the trivial into the vital. This occurs when the diagrams reveal interdepartmental characteristics, characteristics that are beyond the capacity of any one department manager to solve. When this occurs, the problem becomes identified as unique, one that will require in-depth study for solution. The identification of these two characteristics, interdepartmentalism and uniqueness, are the hallmarks of the Vital Few. Hence, redistribution occurs, from trivial to vital. It is often helpful to develop a sequence of Pareto charts in this process, as shown in Figure 10-6.

The Investigation Team Process (Deployment Flow Charting)

If quality management has any single identifying characteristic, it is insistence on the use of statistical methods to solve problems. This insistence derives from the need for precise thinking. The Management Oversight Risk Tree diagrams provide precise guidance in understanding the accident sequence and the safety systems involved in managing that sequence. The Pareto Principle provides precision in establishing what must be worked on first, identification of the vital, in the program of continuous improvement of safety systems.

Precision thinking, by definition, cannot be achieved on the basis of incomplete information. To assure that information is as complete as possible, broad participation in the deliberative processes is necessary. In the accident investigation/analysis process, this has resulted in adopting the team approach to problem identification and solution. To assure precision thinking and response it is helpful to understand the

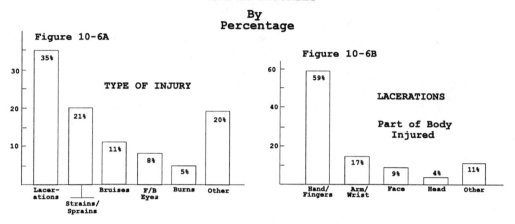
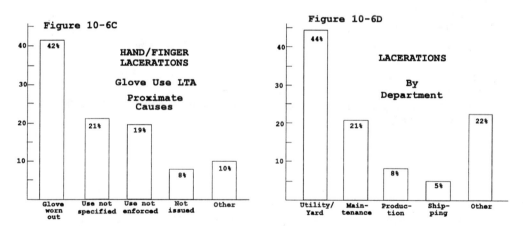

Figure 10–6. Sequence of Pareto charts illustrating percentage incidents of serious/non–serious trauma injuries.

team process. This understanding is the purpose of the technique referred to as Deployment Flow Charting.[10]

The Deployment Flow Chart is based on the familiar flow chart, but one new characteristic has been added, a people coordinate at the top of the chart which indexes the diagram. This "cast of characters" enables identification of the role of each

---

10. This discussion of deployment flow charting is based on the work of Dr. Myron Tribus as presented in the workbook for the videotapes, "Deployment Flow Charting", vols I and II (Los Angeles: Quality and Productivity, Inc., 1989).

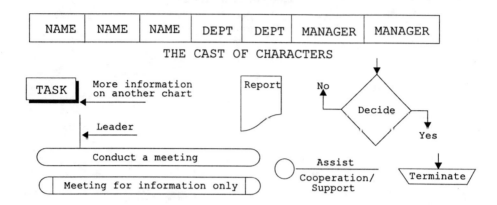

Figure 10–7. Legends for Deployment Flow Charting.

player as it relates to the work of the team. The legends employed in the Deployment Flow Chart are presented in Figure 10–7.

The Deployment Flow Chart for a department team investigation and analysis is relatively simple with no interdepartmental relationships to be concerned about. The chart in Figure 10–8 has been expanded to include the implementation of corrective action and the subsequent monitoring of that action for effectiveness.

This Deployment Flow Chart shows how the people, the processes of accident analysis and the development and implementation of corrective action work together in a simple situation. As the processes become more complex, with, for example, the introduction of interdepartmental or outside influences, the complexity of the cast of characters increases, but so does the value of the chart in achieving precise thinking.

The Cause and Effect Diagram (the Fishbone Diagram)

It would be improper and untruthful to indicate that any one of the tools for precise thinking ranks higher than all others in importance. Each has its purpose, and as different purposes are pursued in the course of deliberations, the specific tool for that purpose becomes individually paramount. As the next phase develops, a new and different tool becomes key.

To define and understand the process or system, the flow chart is the key tool; the Pareto diagram is the key to determining priorities—identifying what to work on first. The next step in the process is to determine the causes that are operative in the problem that the Pareto has selected. This is the purpose of the Cause and Effect diagram.

The Cause and Effect diagram was developed by the late Dr. Kaoru Ishikawa. In 1962 Dr. Deming honored Dr. Ishikawa by calling the chart the Ishikawa Diagram, which remains its proper title in Japan.[11] Because of its characteristic shape it is often

---

11. Kaoru Ishikawa, *What is Total Quality Control? The Japanese Way,* trans. David J. Lu (Englewood Cliffs, NJ: Prentice–Hall, Inc., 1985).

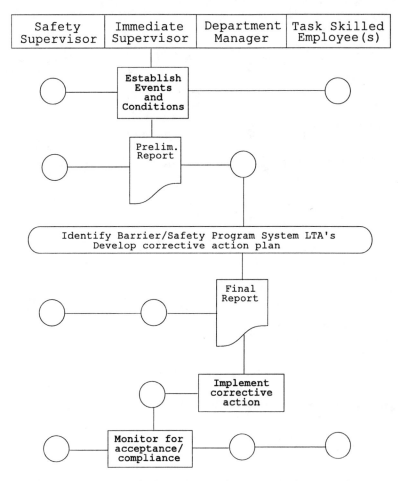

Figure 10–8. Department accident/injury investigation and corrective action compliance/acceptance.

known by its nickname, the Fishbone Diagram. In the diagram the Effect, which is the goal, is located at the right–hand end of a mainstream arrow. The Causes appear as branch streams flowing into the mainstream. The titles of the branches, appearing at the tip of the branch, are the causes. One popular branch identification arrangement is known as the PEEMM method, illustrated in Figure 10–9.

In discussing his diagram, Dr. Ishikawa points out that there are six important attributes to the process involved in its creation and use:[12]

---

12. Kaoru Ishikawa, *Guide to Quality Control*, 2nd ed. (White Plaines, NY: UNIPUB, Kraus International Publications, 1986): 25.

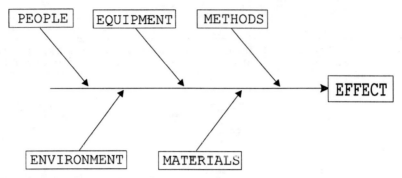

Figure 10–9. PEEMM example of a Cause and Effect Diagram.

1. *Preparing a cause–and–effect diagram is educational in itself:* Get ideas from as many people as possible. Ask everyone. Group discussion brings out stories of experiences and techniques. People learn from each other. All taking part will gain new knowledge. Even people who are new and inexperienced can learn from participating in making a cause– and–effect diagram or from studying a completed one.
2. *A cause–and–effect diagram is a guide for discussion.* Discussions that get off–track are not purposeful. The diagram helps the group focus on the issues; people know what the issues are and how far the discussion has advanced. Complaints, grievances, and pointless repetition can be avoided. Consensus on what action to take is achieved faster.
3. *The causes are sought actively.* The search for causes remains the active motivation. The classification of causes into defined categories facilitates cause identification.
4. *Data must often be collected.*
5. *A cause–and–effect diagram shows the level of understanding.* A diagram that can be drawn up thoroughly shows that the workers doing it know a lot about the process. The more complex the diagram, the more sophisticated the workers are about the process. Beware of simple diagrams with sparse branches. On the other side of the coin, the higher the level of sophistication of the workers, the better the cause–and–effect diagram will be.
6. *The diagram can be used for any problem.* Because this diagram illustrates the relationship between cause and effect in a rational manner, it can be used in any situation; improving product quality and quantity, improving distribution and other systems, improving relationships, including any kind of personnel problem (and most certainly it can be used in the continuous improvement of both safety program systems and management safety systems). The aim is to get results; knowing the relationship between cause and effect will lead to quicker solution.

In his *Guide to Quality Control* Dr. Ishikawa identifies three different types of cause–and–effect diagrams. The first two are Dispersion Analysis and Production Process Classification, both of which are based on finite or discrete data. The third he identified as the Cause Enumeration type. This is the cause–and–effect analysis that applies to the work of safety program management.

In this method, once a desired effect has been selected, all the possible influencing factors are simply enumerated. In compiling the list, everyone's ideas are necessary. For easiest organization of this exercise it is helpful to use a blackboard or flip chart to

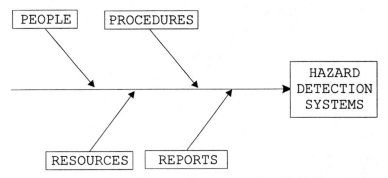

Figure 10–10. Cause Enumeration type of cause–and–effect diagram.

keep the ideas visible. In constructing the list, the causes or influencing factors are organized into causation categories. For example, let us assume that the desired effect is one of improving the system of hazard detection. The categories selected might be (1) People, (2) Procedures, (3) Resources, and (4) Reports. This would produce the basic diagram shown in Figure 10–10.

Whatever work and whatever process is selected, it is not difficult to immediately identify a dozen or more cause factors. As a training exercise for a group of eight prospective Red Team members, a two–fold assignment was made: (1) construct a diagram of the factors they believed to be important in the system of hazard detection, and (2) narrow the factors chosen down to those that either contribute or have the potential to contribute most importantly to the process of detecting hazards. The aim was to identify and prioritize elements of the hazard identification system that were the best candidates for system improvement.

Developing the list and assigning the ideas into categories was essentially a brainstorming process, but it was organized differently. A more structured approach, which Myron Tribus refers to as Nominal Group Technique (NGT) was employed.[13]

The Nominal Group Technique

The issue for the group was stated as "How are hazards detected?" which produced the Effect title of "Hazard Detection Systems". This title was written on the blackboard and kept before the group at all times. A five step process was then followed:

Step One: Working silently, members of the group wrote as many ideas as they could think of on a sheet of paper. The instruction was to express the idea in a "key word" sense, a simple phrase.

Step Two: Upon completion of the lists the leader asked each person in turn to present one idea only. That idea was written on the blackboard **without comment or editing.** The ideas were numbered in sequence as presented. No comments about the idea were permitted. When everyone had an opportunity to speak once, the process was repeated. When a person had no more ideas the response was "I pass". When everyone had passed, the next step was initiated.

Step Three: This is the question and answer period. At this point anyone with a question, by show of hand, could ask a question to clarify what any statement meant.

---

13. Tribus, "Deployment Flow Charting".

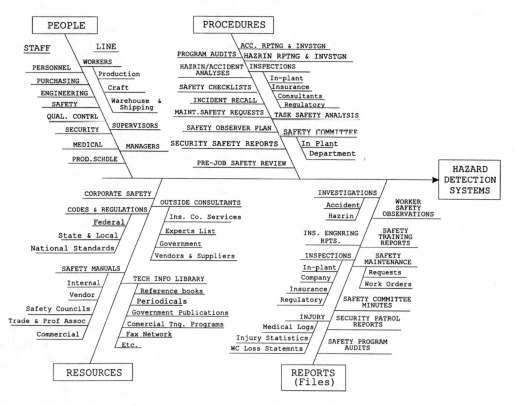

Figure 10–11. Chart of master cause–and–effect (C&E) exercise.

The person who authored the statement was allowed to explain but not justify what was meant. The questioning continued until everyone was satisfied.

Step Four: This is the consolidation and categorization period. The group considered whether or not certain topics overlap, so that some topics might be combined or eliminated. The leader directed this process. Following simplification through consolidation, the group considered the categories into which the ideas could be organized. At this point the thinking was guided by the form of the fishbone diagram, wherein the ideas were grouped into the categories which in the diagram appear as major bones connected to the vertebra. The group was free to select the categories they felt best classified their ideas. The individual elements were shown as minor bones connected to these major bones. The categories selected were: People, Procedures, Reports, and Resources. At this point the information needed to create the master cause–and–effect (C&E) chart was in hand. This chart, constructed later, is shown in Figure 10–11.

In the exercise the categories selected were People, Procedures, Resources and Reports. At this point the thinking was guided by the form of the fishbone diagram, wherein the ideas were grouped into the categories which in the diagram appear as major bones connected to the vertebra. The group was free to select the categories they felt best classified their ideas. The individual elements were shown as minor bones connected to these major bones. The categories selected were: People, Procedures, Reports, and Resources. At this point the in.formation needed to create the master cause–and–effect (C&E) chart was in hand. This chart, constructed later is shown in Figure 10–11.

Step Five: This is the Pareto period, the need now is to simplify the chart by establishing the relative importance of the elements to in order to establish priorities. This is accomplished by "weighting" the elements. The formula for weighting is simple. Each member of the group is given a fixed amount of weights to distribute. That amount is one half of the total number of ideas (N) minus one. The formula is $W = (N - 1)/2$, which is rounded down in the event of a fraction. If there are 22 ideas, W equals 10.

Each person considered the ideas and assigned a weight to each one. The one considered most important was given a weight of 10, the next was given 9, and so on until ten weights had been distributed. The ideas left unscored had a weight of zero. Each person made a list of the weighted ideas to be able to respond quickly in the voting.

A tally of the votes was then taken by the leader. On the blackboard or flip chart the leader recorded each person's weight for each numbered idea, starting with idea number one, then progressing to number two, and so on. Each weight is recorded with a plus sign between entries. (To speed up the tally, a different person was assigned to add up the numbers for each proposal. This is a fun operation; however, if for some reason the tally needs to be secret, the votes can be written on cards which can be collected and shuffled before tallying.)

This process resulted in the data for the construction of four Pareto charts, one for each of the contributing categories. These charts are shown in Figure 10–12. The column heights were determined by the weighted voting score. The bracketed number indicates the number of votes for the category.

The Pareto charts provided the information to enable the construction of a cause–and–effect chart, modified to diagram those elements the group considered most important in the Hazard Detection Systems. That chart is shown in Figure 10–13. Further refinement could be obtained by restricting the voting question to those elements considered most in need of improvement. Since the quality objective is system improvement, this would define the specific systems to work on for improvement of the Hazard Detection Systems, and the selection would have been achieved by consensus.

Dr. Ishikawa points out that the cause–and–effect diagram is a guide to concrete action and that the more use that is made of it the more effective it becomes. The diagram enables the specific thinking required for continuous improvement of safety systems.

The Key Statistical Methods for Specific Thinking

The four diagrams discussed are basically nonmathematical statistical procedures. They qualify as being statistical in nature under the second part of our definition of statistics: *a collection of methods for gathering, analyzing, and drawing conclusions from factual information, a body of theory and methods which help us make wise decisions under conditions of uncertainty.* In statistical terminology these methods are classified as enumerative; they are based on listing or counting, and no discrete or finite measurements or mathematical procedures are involved.

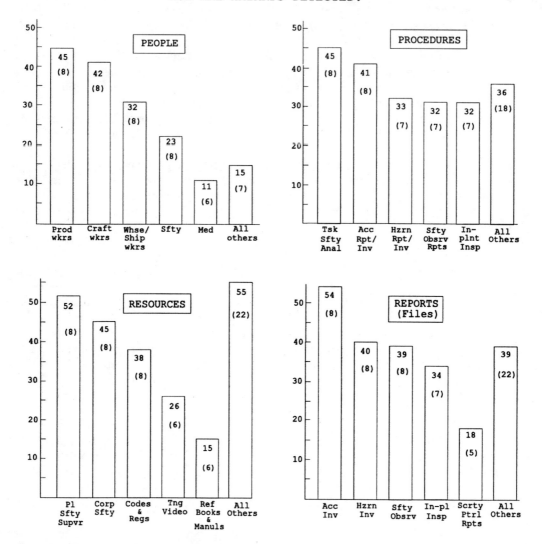

Figure 10–12. Four Pareto charts corresponding to the four contributing categories in the master C&E chart.

The **Flow Chart** developed understanding of the systems or processes that are operative in what is happening. The diagram of The Management Oversight Risk Tree revealed the Accident Triad process of an accident occurrence, and the Anatomy of an Accident expanded the view of the triad to include the safety program factors that directly or indirectly influenced that occurrence. The result is identification of specific safety program elements that are candidates for improvement efforts. The value of the MORT is not restricted to retrospection, to the analysis of an occurrence. It is at least equally valuable for reviewing and evaluating the strengths and weaknesses in the overall safety program, and in serving as a guide for auditing all aspects of the ongo-

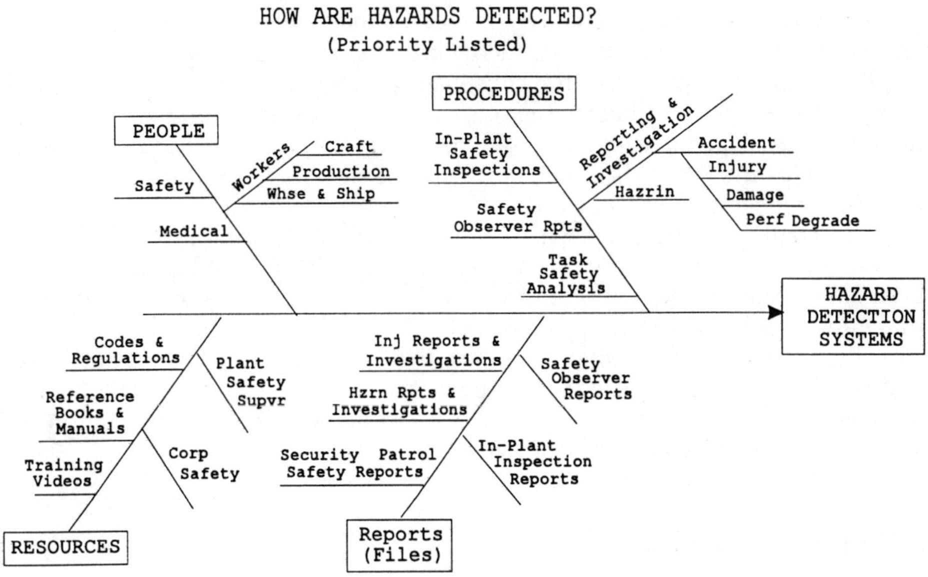

HOW ARE HAZARDS DETECTED?
(Priority Listed)

```
                            PROCEDURES
                                              Reporting &
                                              Investigation
    PEOPLE          Workers  Craft      In-Plant
                           Production   Safety              Accident
                           Whse & Ship  Inspections         Injury
     Safety                                         Hazrin  Damage
                                        Safety              Perf Degrade
     Medical                             Observer Rpts
                                        Task
                                        Safety
                                        Analysis                    HAZARD
    ─────────────────────────────────────────────────────→          DETECTION
                                                                    SYSTEMS
         Codes &            Inj Reports &
         Regulations  Plant Investigations   Safety
                      Safety                 Observer
         Reference    Supvr  Hzrn Rpts &     Reports
         Books &             Investigations
         Manuals
                      Corp   Security Patrol   In-Plant
         Training     Safety Safety Reports    Inspection
         Videos                                Reports

    RESOURCES                          Reports
                                       (Files)
```

Figure 10–13. Master C&E chart modified to diagram those elements the group considered most important in the Hazard Detection Systems.

ing program, both the worksite safety program systems and the management safety systems.

The flow chart is the essential tool for developing appreciation for a system, for the development of the "system's vision" discussed earlier. The flow chart identifies the interdependencies of system components, makes the aim of the system clear, and demonstrates the value of each component to the system.

Systems must be managed to obtain the optimization that benefits all. The statement that an organization is only as good as its people is not true. The organization consists of multiplitive, interdependent systems that reaches maximum value when its value is greater than the sum of its parts. Optimization occurs only when all components win.

The **Pareto Chart** is the first step in making improvements, deciding what to work on first. In making improvements, Ishikawa has pointed out that three things are important: (1) that everyone concerned cooperates; (2) that a concrete goal is selected; and (3) that the improvements have a strong impact. He adds that if all the workers try to make improvements individually with no definite basis for their efforts, a lot of energy will produce few results.[14]

In working toward continual improvement of safety systems, the Pareto diagram presents a clear and distinct view of the relative importance of safety systems' inadequacies, so that all have the same base of common and factual knowledge from which to work. This helps to ensure the cooperation of all and pinpoints the effort on the specific issue, an issue that counts.

---

14. Ishikawa, *Guide to Quality Control*: 44.

The Cause-and-Effect Diagram is the tool that drives and concentrates the cooperative effort. It is the tool that pools the broad knowledge that is required, the ideas of all. It is the essential tool for consensus management, the team approach. The process of creating the diagram is one of "a meeting of the minds", a process that identifies interdependencies and promotes the creation of trust and the removal of the fear of (unjustified) criticism. Both of these are necessary for the cooperation that is essential to the improvement effort.

The Deployment Flow Chart is a tool for defining and organizing the work of system improvement. It is helpful in determining who should be involved, what role each participant should play, and defining the work that needs to be done. It is a flow diagram of the improvement process itself.

These four diagrams—the flow chart, the Pareto chart, the cause and effect chart and the deployment flow chart—are the basic statistical tools for Total Quality Safety Management. They are essential statistical techniques for the optimization of safety program and management safety systems.

Other Statistical Methods
(An Overview of Analytical Statistical Methods)

An interesting analogy of the differences between analytical statistics and enumerative statistics can be drawn in comparing the differences that exist between a pond and a river. Analytical statistics, based on the measurement of finite or discrete values, is a study of the water in the pond. The basic measurement is how much. *The objective of analytical statistics is the detection of variation.* The statistical pictures consist of snapshots. They are static, frozen in time. Variation is detected by comparing the snapshots taken at different times, or stringing the snapshots together in a continuum to represent a flow or trend.

In contrast, enumerative statistics is the study of the flow of the river. The objective is not the detection of variation; variations in the flow are accepted as normal. *The objective of enumerative statistics is the prevention of variations.* The statistical pictures are dynamic; the river is flowing with continual variation even as one speaks. From this point of view, the statistical methods most useful to quality safety management are the enumerative methods discussed above. This, again, is the essence of Dr. Deming's statement about the importance of working on the causes, not the charts. Unfortunately these working methods do not result in impressive charts to present to the boss or post on the bulletin board; where the issue is not methods but results; not how but how much.

It is not unusual to hear the statement, made in jest, that "Accidents make lousy statistics because there are not enough of them." The reference here is not to cause analysis, but to statistical measurement. This is not meant to imply that statistical measurement has no role in quality safety management.

Measurement for results will always be of interest, but it is not where the work is done. The suggested reality is not that there are not enough accidents to be statistically significant, rather, it is testimony to the difficulty of identifying measurable phenomena that would produce statistically reliable data. Counting the defective units in the population, or measuring the variations in physical characteristics by calibrated instruments, results in data that is finite or discrete. Opinion, or variation in observer interpretation of variation definition, cannot introduce or fail to detect variation. Opinion becomes an issue only in the interpretation of the factual data. This characteristic of analytical statistical techniques can be seen as a caveat in the application of those techniques to the phenomena of concern to safety management.

The Run Chart

The simplest analytical statistical tool is probably the Run Chart, a plot of data over time. For safety, an example would be the interval plotting of the Disabling Injury Frequency Rate (DIFR). The result is a trend line. While there is value in knowing what the trend is over the long run, over the short run nothing is revealed. Compare for example the value of a daily charting of the DIFR and the monthly rate, or the annual rate over a period of years.

If the trend is downward we keep on doing what we have been doing. If it is upward the inclination is to work harder at what we have been doing, or to look around for some new gimmick, a booster project, that will hopefully reverse the negative trend. The reality is that the system is stable for the condition it is in, and quick-fix tampering is to be avoided. Over the long run, as system changes for improvement become effective, the Run Chart will indicate whether or not the changes made over the course have resulted in permanent improvement. By itself, however, the Run Chart tells us nothing about the causes of the variations that occur.[15]

The Control Chart

While the Run Chart has been identified as one of the most simple statistical tools, it is the foundation upon which the most sophisticated and meaningful analytical statistical technique for quality management rests. The genesis, development, and significance of the Process Control Chart was discussed in Chapter Three. There the genius of Walter Shewhart was discussed, an intelligence of high order that culminated in the establishment of the cardinal Principles of System Variation. In review, those principles are:

1. Variation in every system is normal.
2. The causes of variation lie either within the system (Common Causes), or outside the system (Special Causes).
3. When a system is running consistently within its upper and lower control limits, under statistical control, and is left untouched, the variations that occur are due to Common Causes.
4. Common Causes arise out of the characteristics of the system which are determined by management, and can only be influenced by management action to improve the system. Workers have no control over common causes.
5. The 85–15 Rule of System Variation: In a normal system, 85% or more of the variations are due to Common Causes. 15% or less are due to Special Causes.

These Principles of Variation and the Rule of System Variation are the foundation upon which the still-evolving new systems of management, including quality safety management, are based. (It is interesting to note that in his latest book *The New Eco-*

---

15. W. Edwards Deming, *The New Economics for Industry, Government, Education* (Cambridge, Mass: Massachusetts Institute of Technology, Center for Advanced Engineering Study, 1986): 35.

nomics, Dr. Deming revised the ratio of system variation to 94% Common Causes and only 6% Special Causes.)[16]

It has been commented that the characteristics of analytical statistical techniques can be seen as a caveat in the application of those techniques to the phenomena of concern to safety management. The sophistication of the control chart increases the need to be mindful of the warning. The difficulties that can arise are indicated by Dr. Deming himself. In *The New Economics,* Dr. Deming presents a flow diagram for use of a control chart.[17] The difficulties are revealed in the five actions that constitute Step One of that flow diagram:

1. Decide what quality characteristic to plot.
2. Decide what kind of chart might be helpful.
3. Decide on a plan for collection of data.
4. Decide scales of format of chart.
5. Achieve statistical control of system of measurement.

The decision on the quality characteristic is the first key to the successful use of the chart. A characteristic that is finite or discrete, a characteristic that can be directly measured or counted, is a basic assumption of the control chart. The validity of the chart is dependent upon the consistency of the characteristic selected for measurement. Variations introduced in the measurement phase would render the exercise invalid.

When measuring the characteristics of the variations in safety system performance, controlling the quality of the data is a problem that must be taken most seriously. That problem is specifically addressed in the fifth action of Step One, achieving statistical control of the system of measurement. What quality characteristic is available that will meet this rigid requirement? What quality characteristic is available that will not present distorting borderline questions? What characteristic is available that will always be seen identically by different reporters, or even by the same reporter at different times and under differing conditions? Statistical control of the data is essential for meaningful measurement. The need for the assistance of a skilled statistician in these efforts is indicated.

For some, the temptation is strong to view accidents as Special Cause phenomena, incidents caused by worker actions that are viewed as fleeting events that arise from outside the system. This temptation derives from the emphasis the traditional paradigm of accident causation placed on the "unsafe act". Slogan exhortations of "Do it right the first time", "zero defects", "error–free performance", etc., are witness to this continuing belief. Observational codes that concentrate on detecting worker discrepant performance for the purpose of enforcement, to the exclusion of the hard work of system improvements, can also be traced to this belief. It is at this point that the stark differences between the traditional accident causation paradigm and the paradigm of system causation of variation come head to head.

In his writings, Dr. Deming emphasizes the point that confusion in this matter of cause identification, first identified by Dr. Shewhart, can lead to costly mistakes. Shewhart identified two mistakes frequently made in attempts to improve results:

> Mistake 1. To react to an outcome as if it came from a Special Cause, when actually it came from a Common Cause of variation.

---

16. Deming, *The New Economics:* 187.
17. Deming, *Out of the Crisis:* 318.

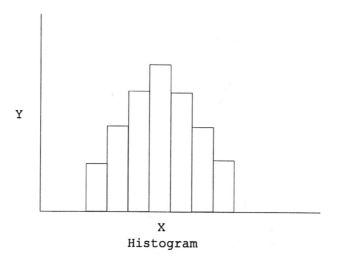

Figure 10–14. Example of a schematic histogram.

Mistake 2. To treat an outcome as if it came from Common Causes of variation, when actually it came from a Special Cause.

Control charts are not difficult to construct, and they are impressive documents for display purposes. But they are highly sophisticated, and, as Dr. Deming cautions, at times even the experts have difficulty in determining their meaning. Such difficulties would be compounded if the data used in construction was suspect in any manner. The wisdom of Dr. Deming's comment on control charts: "Forget about the charts! You must work on the causes" is clear for this author.

The Histogram

In most statistical texts, and subsequently in most introductory academic courses on statistics, the first subject considered is frequency distribution. In the analytical sense this is not a question of *how much* but rather a question of *how often*. The graphic presentation of the *how often* data is called a histogram. *Webster's Unabridged* definition of a histogram is "a graphical representation of a frequency distribution by means of rectangles whose widths represent the class intervals and whose heights represent the corresponding frequencies." Since a histogram is a graph with bars, it is often referred to as a bar chart. A schematic histogram is shown in Figure 10–14.

The histogram can also be constructed as a curve that outlines the pattern created by the rectangles. Characteristically it is a bell–shaped curve that is familiar to all.

Histograms are pictures of dispersion, variation above or below the average for the population being studied. As such, the histogram has an interesting relationship to the process control chart. It is not unusual to see a horizontal histogram added to a control chart to illustrate the spread or dispersion of the data between the upper and lower control limits on either side of the average or $\overline{X}$ line. If the control limits are broad the slope of the histographic line will resemble a gradual mound. As improvement occurs and the control limits narrow to squeeze the histographic line, it will sharply increase in height and narrow in width to more closely resemble a dull pointed obelisk. As this occurs, the $\overline{X}$ line will have declined in value. This is graphic representation of process improvement. Such a histogram is sometimes referred to as

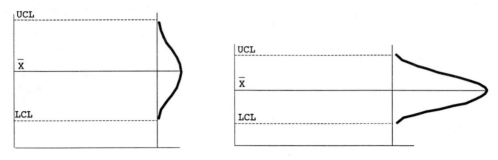

"THE VOICE OF THE SYSTEM"

Figure 10–15. Control charts. Lowering of $\bar{x}$ and narrowing of control limits in the chart on the right indicates process improvement.

"the voice of the system". These relationships are shown schematically in Figure 10–15.

The Scatter (Correlation) Diagram

The scatter diagram is a method of charting the relationship between two variables. All of the analytical methods explored above are methods for handling just one kind of data at a time. Scatter diagrams show the relationship between paired data, telling us how two (or more) variables may be associated and what the degree of the association may be.

The correlation diagram is the tool that tests the degree of association between cause and effect. It does not prove that one variable causes the other, but it does show whether a relationship exists and the strength of that relationship. The significance of correlation was demonstrated in the earlier mention of the Motorola accounting department experience, where it was found that a simple head count of the number of people in the plant was sufficient for costing purposes as opposed to the time–consuming actual measurement of time clock hours. For new employees, a scatter diagram might be used to chart the relationship between training time and the number of defects.

As in all analytical statistical diagrams, the initial step in the construction of a scatter diagram is data collection. For this diagram, however, paired data is collected, 50 to 100 pairs to assure proper representation. In the example of the correlation between the number of people in the plant and the recorded time card hours, 50 to 100 days of head count and time card hours would provide the data. Each day will produce two readings to be plotted. When the data is cause and effect, the cause data is usually placed on the horizontal axis (X) and the effect values on the vertical axis (Y). Since the number of people in the plant directly influences the number of work hours recorded, the head count will appear on the horizontal axis and the work hours on the vertical axis. The resulting plot will result in the pattern shown in Figure 10–17.

The pattern in Figure 10–16 graphically displays the not surprising positive correlation between the two variables. The correlation was judged strong enough to change the procedure from tabulating recorded work hours with the much simpler head count, with substantial saving of clerical time previously unrecognized as essentially

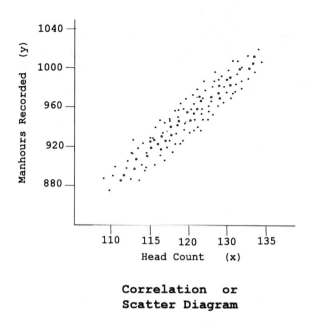

Figure 10–16. Scatter diagram example of two positively correlated variables.

wasted. Figure 10-17 shows the various patterns that Scatter Diagrams can produce and what the different patterns indicate.

1. An increase in Y depends on an increase in X. If X is controlled, Y will naturally be controlled.
2. If X is increased, Y will increase somewhat, but Y seems to be influenced by causes other than X.
3. There is no correlation.
4. An increase in X will cause a tendency for a decrease in Y.
5. An increase in X will cause a decrease in Y. Therefore, as with item 1 above, X may be controlled instead of Y.

In the interpretation of the scatter diagram, a negative relationship is as important as a positive relationship. (In a negative relationship, as Y increases, X decreases, whereas in a positive relationship both increase simultaneously.)

You can only say that X and Y are related. You cannot say that one *causes* another.

The scatter diagram is a highly sophisticated statistical technique, and statistical tests to determine the degrees of relationship are an essential element in the correlation technique. These techniques are beyond the scope of this discussion. Once again, however, the caveat for the control chart and Dr. Deming's instructions in Step One of his flow diagram for use of the control chart apply. However, if statistically valid correlations can be discovered in the phenomena within nonloss or minor loss incidents, or in statistically validated observational codes and injurious or damaging accidents, the correlation diagram would be a valuable technique for quality safety management. Studies directed to this end would seem to have potential for solving the problem of the paucity of data derived from injury–producing accidents. This would enhance the opportunities for statistically valid use of the control chart.

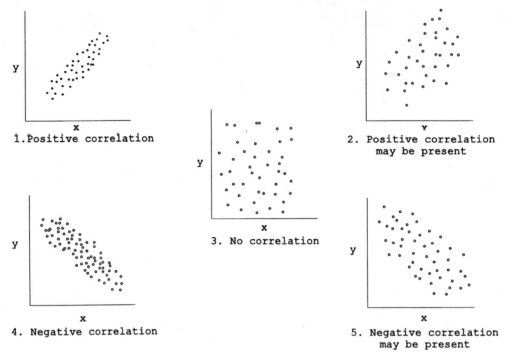

Figure 10–17. Various patterns of scatter diagrams and what the different patterns indicate.

The application of statistical methods to achieve precise thinking is fundamental to quality safety management, no less than it is to all other applications of quality management in the enterprise. They are the technical tools that empower all in the organization, each at his or her own level, to contribute to the objective of continual improvement of the operating systems to the fullest extent of their individual potentials.

11

Employee Involvement in Safety Program Improvements

Worker involvement in safety programs has long been a key point in traditional safety programs. The issue is addressed in the *Accident Prevention Manual for Industrial Operations*, published by the National Safety Council (NSC).[1] In discussing employee responsibilities in the safety program it is pointed out that:

> Employees make the safety and health program work. Well trained employees are the greatest deterrent to damage, injuries, and health problems in the plant or establishment.[2]

Further points are made that employees can observe safety and health rules and regulations, and can work according to standard procedures and practices. They can recognize and report hazardous conditions or unsafe work practices, and develop and practice good habits of hygiene and housekeeping. They can use protective and safety equipment, tools, and machinery properly. They can report all injuries or hazardous exposures as quickly as possible, and, finally, they can help develop safe working procedures and make suggestions for improving existing ones.

The Safety Committee

Workplace safety committees have a history in traditional safety programs extending back in time to the earliest organized programs. In discussing safety and health committees the NSC *Accident Prevention Manual* offers this general description:[3]

> The safety and health committee is a group that aids and advises both management and employees on matters of safety and health pertaining to plant or company operations. In addition, it performs essential monitoring, educational, investigative and evaluative tasks.

Additional comments are offered on the potential for contributions by worker members of joint worker–management committees:

> The joint committee stresses cooperation and a commitment to safety as a shared responsibility. Employees can become actively involved in and make positive contributions… . Their ideas can be translated into actions. The committee serves as a forum for discussing changes… . Employees can

---

1. *Accident Prevention Manual for Industrial Operations, Administration and Programs*, 9th ed. (Chicago, National Safety Council, 1988).
2. Ibid.: 56.
3. Ibid.: 64–65.

communicate problems to management openly and face to face. Information and suggestions can flow both ways... .

Promoting teamwork and worker participation is the ideal for the safety committee. Joint safety and health committees have contributed to the success of many safety and health programs. It is also true that Section 2(b)(13) of the Occupational Safety and Health Act contemplates the possibility of joint safety and health committee initiatives as a supplementary approach to more effectively accomplishing OSHA's objectives.

The question, however, is whether or not the traditional safety committee concept can best serve the new objective of continual improvement of safety program systems. In usual practice, participation on the safety committee is limited; it touches only a few in the organization, even over an extended period of time. It is not an arrangement designed to serve well the objective of broad, active employee involvement in continuous improvement of the safety program. The envisioned responsibilities are general in nature and vaguely stated. The advisory role is revealed in the extensive use of the word "can" in the paragraph commenting on the potential for contribution by worker committee members. There are no imperatives in the charter that these things will happen. And, it can be pointed out that the committee concept is characteristically embedded in the traditional management structure, where the belief that there would be fewer safety problems if only workers would do their jobs correctly is widely held.

The Team Approach

Brian Joiner has pointed out that Frederick Winslow Taylor was the first to bring white–collar people into the factories, mines, and mills to scientifically figure out how to do the work and then tell the blue–collar people what to do.[4] Workers were not to be concerned with problems, that was management's job. In safety, this was the foundation of Heinrich's establishment of the Three E's: Engineering, Education and Enforcement, as the basic elements of organized safety activities. The white–collar **engineers** the design and construction of the plant and scientifically determines what the blue–collars are to do; **educates** them in how it must be done; and then monitors performance and **enforces** compliance with the job instructions.

Joiner makes the point that the Japanese were aware of Taylor's teachings, but they opted to adopt only a portion of his ideas. The Japanese kept the good part of scientific management, the scientific analysis of work, and threw out the bureaucracy and the "white–collar think and blue–collar do". This action was found necessary after several years of experience with management for continuous improvement when it became obvious that the ideas and help, the active involvement, of the people was essential to success.

"Quality Circles" were introduced into the management system in Japan in the 1960s after some ten years of experience with management by statistical methods. The unspoken assumption of Taylor's scientific management had been proven untrue, a handicapping denial of an essential resource. It was recognized that the workers did have heads after all.

---

4. Lloyd Dobyns and Clare Crawford–Mason, *Quality or Else, The Revolution in World Business* (Boston: Houghton Mifflin Company, 1991).

The result has been the evolution of a participative system of management for continual improvement that involves employees at all levels. This system not only allows subordinates, meaningfully renamed associates, to make full use of their experience, knowledge, and abilities, but also relies upon such involvement and contribution. It is a system where decisions on the action(s) to be taken are made by consensus between management and workers. All understand what the problem is, what is to be done and why it is to be done, and all have a vested interest in success. Consensus does not indicate unanimity of opinion, but agreement on the course of action.

Under the philosophy of consensus management this new role for the people doing the work is considered primary. It is regarded as essential to ensure not only consideration of all points of view, but to establish the broadest possible base for ideas on what actions to take, and to assure vested interest in successful implementation. It is a role going far beyond the role of compliance with rules and regulations; of reporting injuries, hazards and unsafe actions; beyond the role of being able to make suggestions. It is a role of active, not passive, contribution. It is a role of responsibility, a role that builds self–esteem, a sense of ownership in the program, pride in the organization, and most importantly perhaps, pride in oneself.

In consensus management worker involvement develops in a natural way. The organizing mechanism is identifying, analyzing and solving problems related to improving the methods and systems that produce the results.

The result is a restructuring of the relationships between workers and management. The manager's job becomes one of helping the people to do the best job possible. This is done by the manager working to improve the system, with the help of the people. In this process the knowledge and insights the workers have accumulated by being on the line, working within the processes and systems day after day, is an invaluable resource.

Although the widespread involvement of the people in the continuous improvement of work was not a feature of the original thinking about quality management, it is now ranked equally important as the use of statistical methods for management by fact. The statistical tools are the key to the identification and analysis of problems; the team approach is the key to developing and implementing solutions to the problems. This is a fundamentally different view of the relationship between workers and management. Peter Drucker has made the point that these developments are changing the social organization of the factory.[5]

Professor Drucker makes the further points that while quality circles were invented and widely used in U.S. industry during World War II they did not survive post–war. In the quality circles fad of the early 1980s, failure was the result. The two developments, the empowerment of the workers and management by the use of statistical methods, are mutually interdependent. This interdependence was not honored in those attempts.

Problems Hidden From Management

Recognition of this interdependency is critical for quality management. This is so because under traditional management most problems are carefully concealed from upper management. This issue was addressed by William B. Smith, Vice President for Quality Assurance for Motorola, a quality managed organization, in an address to the

---

5. Peter F. Drucker, *Managing for the Future, The 1990's and Beyond* (New York: Penguin Books USA Inc., 1992).

1991 Management and Education Conference of the National Private Truck Council. His remarks were directed to "problems hidden from management".

In the Motorola experience it was found that 91% of problems were hidden from general management. The general manager was aware of only 4% of problems on the production floor. General supervision was only a little better off, being aware of 9% of the problems. The largest gap in the information flow appeared between the production supervisors and general supervision. Line supervision was credited with being aware of 74% of the problems. Not surprisingly, the workers were found to be aware of 100% of the problems.

Peter R. Scholtes, in his landmark contribution, *The Team Handbook*,[6] offers the following comment on the changed relationship:

> "Workers and management learn to work together, for Quality Leadership cannot exist where there are adversaries. Managers are still in charge, but they develop a genuine partnership with the workforce. Both sides are better armed with the knowledge and methods to keep the organization in touch with the customer and to provide quality products and services through flawless design, production, and delivery processes. Employees are allowed to make more potent contributions by combining their intimate knowledge of a process with the tools of the scientific approach.
>
> "As they work together to improve quality, workers and management must build mutual respect and trust. The more they help each other employ the scientific approach, the more productivity and quality improve. This environment of teamwork associated with Quality Leadership cannot be developed under Management by Results; conflicting goals, game playing, and distrust get in the way."

While Scholtes is addressing the overall issue of quality and productivity of work, his comments apply to the management and administration of safety programs. However, for the safety program, this role of effective worker involvement will not successfully develop under a management philosophy where there is a widely held belief that there would be fewer problems if workers would only do their jobs properly.

To return to Professor Drucker's point on the interdependency of employee involvement and the use of statistical methods, it can be noted that these two fundamentals of quality management are mentioned in the analytical logic of the Management Oversight Risk Tree. However, since the MORT is a product of pre–quality management philosophies, neither are given prominent attention.

Employee involvement is mentioned in the discussion of performing Task Safety Analysis shown in Figure 7–17, under the heading of "Employee Suggestions and Inputs". A review diagram of the analytical logical path is presented in Figure 11–1.

This meager mention of employee involvement is testimony to the MORT origins in traditional management thinking. This cannot be regarded as an omission or oversight by MORT's creators. It is simply testimony to the need to continually re-examine and upgrade the analytical logic system, to continually improve the Management Oversight Risk Tree, to adapt it to the changes that occur in thinking or methodology.

The limited mention of employee involvement might be regarded as an omen. Professor Drucker has pointed out that even the most successful practitioners of the

---

6. Peter R. Scholtes, *The Team Handbook, How to Use Teams to Improve Quality*, (Madison, WI: Joiner Associates, 1990): 1–10.

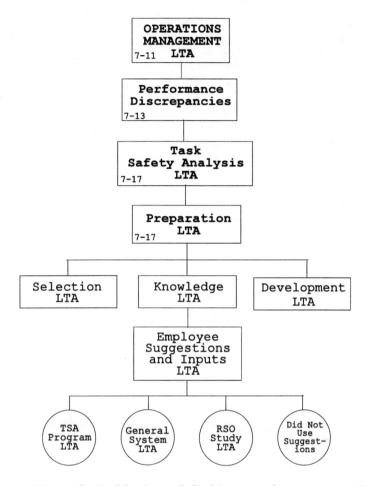

Figure 11–1. MORT analytical logic path linking employee suggestions and inputs to operations management.

precepts of statistical quality control (SQC) do not thoroughly understand what it really does. "Generally it is considered a production tool. Actually, its greatest impact is on the factory's social organization."[7]

The social structure of traditional management is based on Taylor's "white–collar think—blue–collar do" stereotype. Under the precepts of total quality management the manager's job has been redefined: The people work *in the system,* the job of the manager is to *work on the system,* to improve it, *with the people's help.*

The new order might be phrased as "blue–collar do and think—white–collar study, coach, and teach". This is not only fundamental change in function; it produces the fundamental change in social relations to which Peter Drucker refers. The "boss" is no longer; he or she becomes the mentor of the team. The "worker" is no longer; he or she becomes a player on the team, and may indeed serve as leader of the team. All

---

7. Drucker: 302.

contribute to the process of improvement to the extent of their individual capabilities. This is the significance of the "team approach".

These changed relationships provide the content for the omen seen in the meager mention of employee involvement in the MORT analytical logic. Drucker has commented that the effect of these changes has proven to be a "...sharp drop in the number of non-operators: inspectors above all, but also the people who do not do but fix, like repair crews and 'fire fighters' of all kinds. ...moreover, first-line supervisors are also being gradually eliminated, with only a handful of trainers taking their place."[8]

What is the portent of the omen for safety management? What are the implications of the quality mandate for maximum employee involvement, for the Team Approach, in the tactics and strategies of the safety program? Are these developments a threat or a boon to safety practitioners as staff contributors to the organizations efforts?

The suspicion is that individual answers to these questions will vary widely, as widely as different safety professionals would vary in their answer to the question "What is safety all about?" For the writer, however, the answer is clear. Active and widespread employee involvement in all aspects of the safety program is a development long past due in coming. This opinion is offered, however, in full recognition of the fact that fundamental and often difficult changes in thinking and practices on the part of safety professionals, on the part of both supervisory and general management, and on the part of the people, are imperative.

The "Systems" Vision

The first imperative is to develop the ability to visualize that all things that happen, the way things work, have a basis in a system. The system may be well defined and understood, only vaguely defined and poorly understood, or not defined or recognized at all. Large systems are a complex of practically endless mini-systems (the MORT system of safety programs, for example). And the total enterprise consists of almost countless large systems.

Within the enterprise it is inevitable that conflicts between systems will arise. Consider, for example, the failure of a skilled, safety trained mechanic in an auto assembly plant who was killed as the result of a failure to lock out the equipment he was working on, a clear violation of a cardinal and OSHA required safety procedure.[9] The difficulty was that, for the company, this was the third fatality from this same cause in the last two years. The verdict was unanimous as to the cause. Management, including the safety staff, attributed all three incidents to human error and the corrective action was based on that finding. Bulletins were issued and meetings held company-wide to emphasize again the absolute need to follow the lock-out procedure.

The suggestion is that the failure of *three* skilled, safety trained mechanics to follow the requirements of the safety system was not a human failure but a human response to another, and in this case conflicting, system; the perceived requirement in the maintenance system for quick emergency repair response. That system, a system

---

8. Drucker: 303.

9. This example arose in a conversation with a neighbor, a trouble-shooting manufacturing engineer for a large American auto manufacturer. His reaction to the comment that the basic cause of 85% of accidents is not unsafe acts but oversights and omissions in the management system was immediate and heated, and he cited the experiences related here as proof that "human error" is the culprit.

tied directly to each mechanic's career expectations, was much more in mind than the demands of the safety system.

There is little doubt that the safety staff and the management felt they had a sound procedure in place, and that all involved were trained in its application and imperative. At the same time the maintenance managers were no doubt comfortable with their response to upper management concern for expeditious emergency repair response. The conflict between the two systems, unrecognized by both maintenance and safety, however was, in all likelihood, fully recognized by the workers.

The development and adoption of a systems perspective is a primary requisite of total quality safety management. That perspective must include recognition of the fact that all systems, large and small, continuously vary in their operation, and it is in those variations that opportunities for improvement will be found. Opportunities for improvement will also be found in the areas of intersystem conflict, as demonstrated in the lock–out incidents.

The real experts on system shortcomings and conflicts are the workers. They live with them on a daily basis and suffer the problems and contradictions they present in silence. While tactical suggestions on improving safety may be encouraged, suggestions on the strategies employed in how things should be managed have rarely been solicited. Few bosses warmly embrace the idea of having employees tell them how to run their department or program. A team investigation and analysis of the first lock–out fatality, involving skilled task personnel, production, maintenance, safety, and even others, employing statistical methods of analysis, would have been an approach with great promise. Team development of a cause and effect analysis would develop system definitions that in every likelihood would have revealed the intersystems conflicts. The more experience and points of view included in the deliberations, the greater the chances for more complete understanding.

The Safety Capabilities of All

The second imperative of the quality mandate for effective and broad employee involvement impacts the matter of safety training. It has been commented that in the new system "all contribute to the extent of their capabilities". For safety management the corollary responsibility of this mandate is continual improvement of the safety capabilities of all.

Organizations that become involved in adopting and practicing continual improvement management and the team approach quickly discover that the single most time consuming, expensive, and identifying feature of the effort is training. Training and education for everyone constitute two of Dr. Deming's Fourteen Points for Management, the centerpiece of his program for improvement of quality, productivity, and competitive position. "Help the people to improve. I mean everybody" is his comment.[10] The aim, of course, is continual improvement in the capabilities of all the people.

Safety management is knowledge work. This is recognized in the Management Oversight Risk Tree with the identification of "Technical Information" as the first principal safety program system in the diagram. It is from this base that all else flows. The base is monumental, both in size and scope; it is, in fact, so monumental that no one individual could conceivably become proficient in every aspect of the subject. So,

---

10. Mary Walton, *The Deming Management Method* (New York: Dodd, Mead Company, 1986): 85.

as in other great knowledge areas, specialization becomes the answer. As someone has said "We come to know more and more about less and less." However, the beneficial result has been that safety training programs on just about every conceivable aspect of "how not to be injured" are available.

The difficulty is that the majority of these safety training programs are simply that, training; training in techniques, training in answers prepared by experts on what or what not to do under the circumstances being discussed.

And, since the underlying belief that most of the trouble is people–caused remains strong, the programs are slanted to the need for behavioral patterns that will avoid the trouble. Such programs do little to increase the safety capabilities of the people. They do not address the basic question, "What is safety all about?" Increasing safety capabilities requires increased understanding to enable thinking. Safety rote training does not increase understanding. Understanding requires safety education.

It is true, and it always will be true, that safe behavior is essential to safe performance. But, the point of this discussion is improving safety capabilities, not the point of instilling and enforcing safe habits. It is a matter of respecting the capabilities of the people and helping those capabilities to grow.

What form should employee safety education take? What curriculum is needed? Our fundamental question "What is safety all about?" lies at the heart of these questions. The ultimate objective is to have *all* in the organization able to give a clear answer to that question and with the same answer. The words might vary, but universal understanding of the essential elements of accident causation and the safety program systems of control and prevention is a fundamental requirement. The safety education program therefore must include the education of management in the same curriculum. And very serious consideration should be given to joint educational sessions to build understanding and acceptance of the team method and consensus decisions in problem analysis and solution.

The basic elements of a revised safety education program have been set forth in earlier chapters. The Accident Triad, and the Anatomy of an Accident, developed from the Management Oversight Risk Tree, established the new paradigm. It is upon the foundation of that new safety paradigm that the structure of total quality safety programs will rest.

The day–to–day living quarters in the structure will be the area of safety tactics, the use of barrier systems to control the results of unwanted energy transfers or environmental conditions. This will constitute the first floor resting on the foundation. And it is here that the most immediate, and perhaps the largest, potential for employee involvement in building the edifice rests. Increased appreciation for the role of energy in all its variety of forms in accident causation must be instilled to become the reactive way of viewing any accident. "It's just one of those things that happens sometimes" should never be heard again.

Increasing the safety capabilities of all, the new objective of safety education, will not only strongly advance on–the–job safety, it will carry over into the routines and activities of daily living, off–the–job safety. And as the capabilities to control the results of unwanted energy transfers or environmental conditions continue to advance, the interest and ability to contribute to the strategies of safety, the systems and programs to prevent damaging incidents in the first place, will increase as well. The contributions of William Haddon, Jr. and others as presented in Chapter 6, "The Concepts of Energy and Barriers," provide a potential starting point for the improvement of the safety capabilities of all.

The Economic Imperative

One of the primary imperatives of quality management is the elimination of wastes; the waste of assets, the waste of materials, and the waste of the time of people, all people, all levels. Toleration of all of these forms of waste might be said to be a hallmark of traditional American management, a distinguishing and identifying characteristic of the system.

> *Continuous improvement.* For the company and its employees, continuous improvement means that the very first day a process operates must also be the first day that everyone works to improve that process. This means eliminating waste that is, searching out and removing any activity or resource that is not essential in producing the company's products and services. (These activities are also called non–value activities.)

This statement by William Turk,[11] director of cost accounting at Harley–Davidson, Inc., sets forth the economic imperative for the elimination of waste through the active involvement of every employee in that process. The model established here is not unique to Harley–Davidson, it is characteristic of quality managed organizations. It has particular significance for the safety program.

From the earliest days of organized safety programs, safety practitioners have been diligent in their efforts to alert management to the wastefulness of accidents without great success. The difficulty, as discussed earlier, was rooted in the assumptions and mechanisms of the standard cost accounting systems, which built the wastes of accidents into the system. Those days are finally coming to a close as accounting practices are being revolutionized under the severe requirements of quality management.

The driving force of traditional accounting systems is the concept of direct labor costing as the prime element of control. The method is directed to the control of all three elements of the accounting system (inventory control, product cost, and factory performance) with one number—direct labor. In quality managed organizations, as manufacturing operations improved and wastes were being eliminated, the traditional direct labor performance measures indicated deterioration, although all could see the improvements being made. It became increasingly clear that the accounting system was no longer providing effective and useful information. It was, in fact, providing misleading information. The conclusion was that continuous improvement applied to accounting systems as well as production systems.

The result has been an emerging system that some have called "activity based accounting". In the new system it has been found that group measurements of operations are better than individual measurements, and that nonmonetary measurements may be more appropriate, quicker and easier to access, and more readily understood than monetary measures. One of the keys to these recognitions was the identification of relationships previously unrecognized. For example, at Harley–Davidson it was found that what was really relevant to spending on labor was not whether the people in the plant worked 4000 or 4050 hours a day, but the simple fact of how many people were in the plant.

The implications for safety in the development of activity based accounting is that the losses associated with accidents and incidents of unwanted energy transfers will

---

11. William T. Turk, "Management Accounting Revitalized; The Harley–Davidson Experience," *Cost Management*, Winter 1990. This statement and much of the following discussion is based on Mr. Turk's valuable article.

be identified for what they are when they occur; wastes of assets, wastes of materials, and wastes of the time of people. In this manner the economic role of safety and incident/accident prevention will be recognized. The result will be a management imperative that these wastes be controlled and eliminated. Finally, safety's familiar "iceberg" is being placed in dry–dock for all to see in its entirety. The development introduces a measure of the cost effectiveness of the safety program not previously available.

The Information Imperative

In striving for cost effectiveness against this standard of the elimination of wastes that arise out of unwanted energy transfer incidents, the help of the people is essential.

Recall the Motorola experience with "problems hidden from management". Whether the figures presented are precisely accurate or not makes little difference. The general magnitude for each category is all that matters. It also matters little whether the concealments are intentional omissions, or simply oversights. The result is the same—wastes that are tolerated and built into the system. In either event, it is the people who are most aware of what is actually occurring.

One absolute prerequisite for effective employee involvement in these waste containment efforts is the removal of fear of criticism or reprimand for what has occurred. Establishment of the new paradigm of accident causation, which attributes 85% or more of accident incidents to system oversights and omissions, is therefore an imperative for successful employee involvement on the scale that is necessary. Only when all the people know that they will not be criticized or reprimanded for things over which they had no control will the open and honest cooperation that is required be forthcoming. The program of Total Quality Safety Management will succeed with nothing less.

And there is a further information imperative involved in the economic mandate for the elimination of wastes. The traditional major emphasis on protecting the people from injuries and hurt must be broadened to include the waste of assets, the wastes of materials, and the wastes of degraded performance from the same phenomena.

This is not a new concept. Protection against financial losses due to "accidents" to property has long been a major concern. It is the original risk that nurtured the evolution of the principles of insurance. The development of those principles and that industry can be seen as in response to Pareto's "Vital Few". The void that has occurred rests not in the treatment of catastrophic wastes, but in the unwanted energy transfers that occur in ongoing operations that produce eroding wastes of assets, materials, and the people's time. These eroding incidents constitute the "Trivial Many" of Pareto's analysis. In the past the financial losses from these wastes have been buried in the standard cost accounting systems, unidentified and tolerated because they were viewed as affordable.

The contributions of Frank Bird, Jr., and others in this area of loss prevention are recognized, but unfortunately the response to their findings and recommendations has not been overwhelming. By and large, systematic reporting and investigation of "no injury" incidents that are not catastrophic does not take place. Yet they occur with frequencies that greatly exceed that of injury–producing incidents.

Many of these incidents have minor significance, except that of frequency, while others are incidents with significant potential for injury or damage. While it is recognized that some safety programs do respond to these incidents through programs of

Reported Significant Observation (RSO) systems, such programs exist primarily in operations that involve very high energy levels. Most are content with the fact that the wastes of such incidents will simply be absorbed, unidentified, in the accounting system. This will change as activity based accounting replaces standard costs.

However, this is only one imperative for expanding the scope of the safety program to include the routine wastes of materials and assets through unwanted hazard related incidents. The second imperative is technical, and it centers on the problem of the significant statistical information that is needed to guide continuous improvement of the safety program systems.

One major problem that safety has with numerical statistics rests in the fact that there are not enough injury incidents to form a significant base for analysis. Statisticians would classify the problem as a population that is too small to yield reliable results. Broadening the statistical base through the development of a manageable system of incident reporting is a challenge that must be met to serve the statistical needs of the continuous improvement imperative. Such expansion might very well lead to significant correlation analyses such as that experienced by the Motorola accountants.

The third imperative for the broadening of the program, the information imperative, may be the strongest imperative of all for direct employee involvement in safety program management. The employees represent the source of the information that is needed. As the Motorola experience revealed, they are aware of 100% of the problems. Under most traditional management directed safety programs they feel they have nowhere to go with their information. Offering suggestions to the boss is not the thing to do. They may be reluctant for reasons of fear of rejection of the idea, being viewed as a complainer, or criticism by fellow workers. Finally, they realize that nobody is really interested anyhow.

But the information issue is much greater than simply awareness of problems. There is strong evidence developing that databases and computer networks are revolutionizing information handling throughout many organizations. The developments impact strongly on Drucker's point about the changing social hierarchy. A special report in *Business Week* magazine[12] has provided the following quotes:

> Corporations have been laying off large numbers of middle managers because re–engineering and technology make it possible to do without them. In the old corporate hierarchies, middle management's function was to transmit information from the field or factory to the executive suite and relay commands from the corner office back to the troops. Data bases and computer networks now do the job—faster, better, and for less. "People who don't add value are going to be in trouble," says Melvyn E. Bergstein, president of Technology Solutions Co., a systems integrator, "If your job is just passing orders along, you could get lost in the shuffle."
>
> "In factories, people on the line are thinking like industrial engineers," says Richard L. Florida, professor of management at Carnegie Mellon University.
>
> GE's big breakthrough has been giving workers flexibility and unprecedented authority to decide how to do their work, "All of the good ideas—all of them—come from the hourly workers," says Cary Reiner, GE's vice–president for business development.

---

12. "The Technology Payoff," *Business Week*, No. 3323, June 14, 1993: 57.

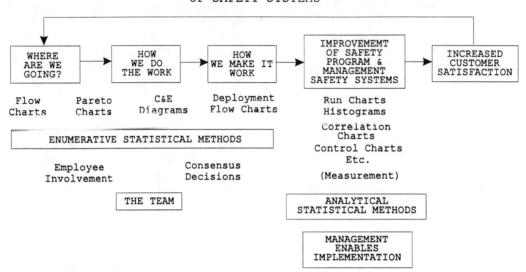

Figure 11–2. Author's vision for the total quality methodology of continual improvement of safety systems.

Improved productivity is still the only way we know to boost a nation's standard of living. And U.S. productivity gains, in turn, can come only from technology and the workplace changes it spawns. "It's absolutely unambiguous," says Columbia University economist Frank R. Lichtenberg. "Productivity gains are crucial for our long-term economic well-being."

The *Business Week* special report, which is not unique in current business literature, makes clear that revolutionary changes in the structure of American business management are taking place. That restructuring centers entirely on the matter of developing and handling information. Safety and accident prevention is a knowledge based function. For safety management the implications of the knowledge revolution that is occurring at an ever-increasing pace are difficult to predict. One thing that is clear is that employee involvement and participation in the active management of safety programs is not only essential, it is inevitable. The task of the safety professional will be to assist and add value to these changes. That is both the challenge and the opportunity. The key lies in education, not training. It lies in increasing the safety capabilities of all and organizing to capitalize on those increased capabilities.

Summary: the New Methodologies

It has been pointed out that the flow chart is the statistical tool that defines the system. Figure 11–2 is the author's vision for the total quality methodology of continual improvement of safety systems.

12

TRANSFORM: To Change Completely or Essentially in Composition or Structure[1]

The Deming theory of management, continuous improvement of the systems of production and service to stay in business and create jobs, requires not adaptation of traditional management principles and practices, but their replacement with new systems of principles and practices. This is the awesome challenge of Dr. Deming's call for the transformation of traditional American management methods.

In traditional management, decisions are dominated by relationships of power. The result is relationships that are competitive and myopic to the realities of interdependence. Too often the result is maximization of the seemingly unrelated subsystem at the expense of other subsystems, and eventually the system as a whole. The logic of competition is win–lose logic; for a win to occur, the risk of corresponding unforeseen loss elsewhere in the system is great enough to be almost inevitable. Dr. Deming's call for transformation is a call to replace win–lose logic with win–win logic—principles of reasoning that require that a change in any subsystem be beneficial to the entire system, or at least cause no loss elsewhere in the system. His call is to replace relationships that are competitive with relationships that are cooperative.

The System of Profound Knowledge[2]

Dr. Deming makes the point that the first step in the change process is transformation of the individual. And the first step in the transformation of the individual is understanding the system of profound knowledge. His presentation of the system of profound knowledge consists of four interrelated parts:

- Appreciation for a system
- Knowledge about variation
- The theory of knowledge
- Human psychology

Appreciation for a System

"A system is a network of interdependent components that work together to accomplish an aim. Without an aim there is no system. For optimization,

---

1. *Webster's Third New International Dictionary of the English Language* (Springfield, MA: Merriam–Webster Inc., 1986).
2. W. Edwards Deming, *The New Economics for Industry, Government, Education* (Cambridge, MA: MIT–CAES, 1993): Chapter Four.

a system must be managed. Optimization is a process of coordinating and concentrating the efforts of the components on the achievement of the aim."

Knowledge about Variation

"Variation is a constant in life. It is a constant in the operation of every system, and it is in the variations that the information essential for system improvement rests.

"System variations arise from causes that lie both within the system (common causes) and outside of the system (special causes).

"A system that is subject only to common cause variations is stable, which renders variation predictable. A system that is subject to both common and special cause variations is unstable, which renders future performance unpredictable. Confusion between the two conditions is an invitation to disaster."

The Theory of Knowledge[3]

"Knowledge requires theory; in the absence of theory all one has is information. Data is not knowledge. Knowledge enables rational prediction. A statement that conveys knowledge predicts the future outcome with the risk of being wrong, and it must fit observations from the past without failure. Even a single unexplained failure of a theory requires modification or even abandonment of the theory. Management in any form is prediction."

A flow diagram of the Theory of Knowledge is presented in Figure 12–1.

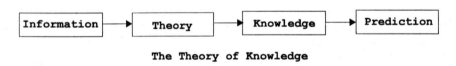

The Theory of Knowledge

Figure 12–1. Flow diagram of the Theory of Knowledge.

Human Psychology

"The requirement for the understanding and employment of human psychology brings the entire matter of personal interrelationships into the system of profound knowledge. This can be seen as the focal point for the change from relationships that are power based (competitive), to relationships that are cooperative: "In quality management the managers functional role is tightly defined: The job of the manager is to work on the system, to improve it with the workers' help."[4]

---

3. Deming, *The New Economics for Industry, Government, Education:* 104–105.
4. Ibid.: Chapter One.

"The significance of this redefinition of the role of managers is their assignment to a role of leadership. In this new role, managers becomes the suppliers of services to their customers, the services of leadership. To be successful in this relationship, leaders must not only develop understanding of the mental, attitudinal, motivational or behavioral characteristics of individuals and groups of individuals, they need to develop an understanding of the psychology of leadership as well, its obligations, principles, and methods."

The system of profound knowledge establishes the new mindset that is essential for accomplishing this transformation, a new landscape for thinking. They are topics that have been absent from the training and educational agendas of Western managers. They are profound in the sense that they are essential deep learning and insight for comprehension of the new systems of management. They provide the methodology for the development of the new systems. Deming makes the point that understanding the system of profound knowledge will provide the individual with new perceptions for events, numbers, and the interactions between people.[5] These new perceptions provide not only a new basis for management, but a new basis for leadership.

An apt phrase for the new basis of leadership might be "help the people". It is the job of the leader to select well, to train well, and to educate for understanding through counseling; to lead on a daily basis, creating a team feeling, in order to work toward continuous improvement. Leaders must show their need and appreciation for their help with ideas for improvements that will make jobs easier, safer, more interesting, etc., to nurture their sense of pride in their contributions and self–esteem.

The reader will recognize at least two of the four elements of profound knowledge as the basis for the discussions and materials presented in the preceding chapters. Appreciation for a system and knowledge of variation have received more attention than the elements of the theory of knowledge and human psychology, but in practice, all four elements are equally important and totally interdependent. Gogue and Tribus make the point that the system of profound knowledge should not be compartmentalized in understanding, and that the ability to use all four elements of the system at once is what distinguishes a true manager from the narrow specialist.[6]

The Fourteen Points for Management[7,8]

While the system of profound knowledge establishes the mental landscape for continuous improvement of quality, productivity, and competitive position, Dr. Deming's Fourteen Points for Management provide the action method, the rational plan for continuous improvement. They are procedures that establish both the aim and criteria for measuring performance. He refers to them as "Principles for Transformation for Western Management".

---

5. Deming, *The New Economics for Industry, Government, Education*: 95.
6. Jean–Marie Gogue and Myron Tribus, "How to Update the BOSS," *Journal for Quality and Participation*, Association for Quality and Participation, July/August 1993: 10.
7. Deming, *Out of the Crisis* (Cambridge, MA: MIT–CAES, 1986): Chapter Two.
8. William S. Sherkenback, *The Deming Route to Quality and Productivity, Road Maps and Roadblocks* (Washington: CEEPress Books, George Washington University, 1992).

As presented by both Dr. Deming in his book *Out of the Crisis,* and William Sherkenbach, a principal interpreter of the Fourteen Points, in his book *The Deming Route to Quality and Productivity, Road Maps and Roadblocks,* the exposition of the principles is understandably generally couched in language and examples that come from processes that produce recognizable results of some kind, either finite products or services that directly produce recognizable results. For most students seeking understanding that will enable application of the principles, the presentations are perfectly adequate. For others, however, translation into practice is more difficult. This seems to be particularly true for those in staff functions that are advisory in nature, those functions where the translation of information and knowledge into practices is dependent upon action by others. However, this difficulty of translation not withstanding, the Principles are intended to apply to all.

In the following presentation of the Fourteen Points, the objective is translating their intent and meaning for the management of safety program systems and management safety systems. The first step in this effort is presenting the principles in the order they are presented by Dr. Deming:

1. **Create constancy of purpose** toward improvement of product and service, with the aim to become competitive, stay in business, and provide jobs.
2. **Adopt the new philosophy.** We are in a new economic age. Western management must awaken to the challenge, must learn their responsibilities, and take on leadership for change.
3. **Cease dependence on mass inspection to achieve quality.** Eliminate the need for inspection on a mass basis by building quality into the product in the first place.
4. **End the practice of awarding business on the basis of price tag.** Instead, minimize total cost. Move toward a single supplier for any one item, on a long–term relationship of loyalty and trust.
5. **Improve constantly and forever the system of production and service,** to improve quality and productivity, and thus constantly decrease costs.
6. **Institute training on the job.**
7. **Institute leadership.** The aim of supervision should be to help the people, machines and gadgets to do a better job. Supervision of management is in need of overhaul, as well as supervision of production workers.
8. **Drive out fear** so that everyone may work effectively for the company.
9. **Break down barriers between departments.** People in research, design, sales, and production must work as a team, to foresee problems of production and in use that may be encountered with the product or service.
10. **Eliminate slogans, exhortations, and targets for the work force asking for zero defects and new levels of productivity.** Such exhortations only create adversarial relationships, as the bulk of the causes of low quality and productivity belong to the system and thus lie beyond the power of the work force.
11. **Eliminate work standards** (quotas) on the factory floor. Substitute leadership. Eliminate management by objective, management by the numbers and numerical goals. Substitute leadership.
12.(a) **Remove barriers that rob hourly workers of their right to pride of workmanship.** The responsibility of supervisors must be changed from sheer numbers to quality.
12.(b) **Remove barriers that rob people in management and engineering of their right to pride of workmanship.** This means, inter alia, abolishment of the annual or merit rating and of management by objective.

Point Group Title	START focusing on improvement of process	STOP focusing on judgement of results
PURPOSE	1. Create constancy of purpose. 14. Put everybody to work to accomplish the transformation	
LEADERSHIP	7. Institute leadership.	11. Eliminate numerical goals and quotas 12. Remove barriers to pride of work. 8. Drive out fear.
COOPERATION	2. Adopt the new philosophy	9. Break down barriers between departments. 4. End the practice of awarding business on the basis of price tag.
TRAINING & EDUCATION	6. Institute training on the job. 13. Institute a vigorous program of education and self-improvement.	
IMPROVEMENT OF PROCESS	5. Improve constantly and forever the system of production and services.	3. Cease dependence on mass inspection. 10. Eliminate slogans and exhortations.

Figure 12–2. Table showing the classification of Deming's Fourteen Points by Tveite.

13. **Institute a vigorous program of education and self-improvement.**
14. **Put everybody to work to accomplish the transformation.** The transformation is everybody's job.

Classification of the Fourteen Points

This presentation of the Fourteen Points may strike some as an unorganized list, difficult to follow. There are, however, themes discernible within the list. These have been helpfully identified in a paper by Michael D. Tveite.[9]

Included in his analysis is grouping the points by titles that characterize their nature. The titles he has developed are: Purpose, Leadership, Cooperation, Training and Education, and Improvement of Process. The resulting table is shown in Figure 12–2.

---

9. Michael D. Tveite, *The Theory Behind the Fourteen Points: Management Focused on Improvement Instead of Judgement* (Minneapolis: Processs Management International, 1990).

Dr. Deming presents the Fourteen Points as principles for transformation, universally applicable to the process of continuous improvement. Each statement has been carefully worded as an expression of how things will be better for all concerned. Each statement is both an aim and a standard for measuring progress toward that aim. It is intended that the aims will be achieved through the application of the system of profound knowledge.

The principles are unranked. For many, at first introduction, some particular principles may appear to have no application to their situation, or their application will be beyond the scope of their authority or sphere of influence. Application of the Pareto Principle will resolve any problems this may cause. Ranking—deciding what to work on first—is left to the practitioner.

In the review that follows, it will be noted that the remarks for some of the principles are extended, while others are rather brief. This reflects the author's "feeling" about the relative importance of different points to quality safety management. Others may have different priorities based on their different situations and experiences as safety managers. This is as it should be. The author's remarks are not intended to serve as a substitute for the study of the Fourteen Points as presented by Dr. Deming, Sherkenbach, and other authorities.

For the most part this review follows the pattern shown in Figure 12–2.

PURPOSE: Start focusing on improvement of process.

POINT 1. CREATE CONSTANCY OF PURPOSE TOWARD IMPROVEMENT OF PRODUCT AND SERVICE, WITH THE AIM TO BECOME COMPETITIVE, STAY IN BUSINESS, AND CREATE JOBS.

Dr. Deming repeatedly reminds you that doing your best on the job is not good enough. "You have to know what to do. Then do your best." This is the message of constancy of purpose—knowing what to do. Once that is established then do your best—maintain consistency of purpose.

How does a statement of constancy of purpose help safety managers understand what their jobs are? The statement of constancy of purpose establishes the direction and content of the job, and provides a standard for continually evaluating the consistency of purpose in job activities.

Statement of Constancy of Purpose for Safety Management Professionals

The purpose of safety program management activities are:
1. To add value to work and the company's products and services through the elimination of the wastes that arise out of accidents (the customer–driven component).
2. To assist in the development and establishment of effective safety program systems and management safety systems that nurture pride of accomplishment and self–esteem for all (the team–fueled component).
3. To continually improve all safety systems (the even–keeled approach).

The economic role of safety program activities is the elimination of wastes that arise from hazard related incidents. The wastes that accrue to HAZRINs diminish the productivity of the system, whether they are covered by budgetable insurance or stan-

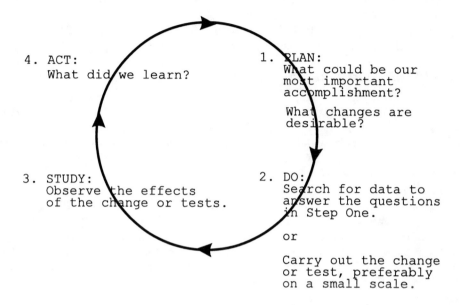

STEP 5: Repeat Step 1 with what was learned.
STEP 6: Repeat Step 2 and onward.

Figure 12-3. The Shewhart Cycle.

dard cost accounting charges or not. These wastes are ultimately borne by the customer and society.

The elimination of wastes due to hazard related incidents is the operational definition of safety program systems, a definition that communicates the concept of safety program activities, a definition that people can do business with.

The statement of constancy of purpose sets the course today for tomorrow. In terms of Dr. Deming's reminder about doing your best, the statement tells us what we must be doing so that we can meaningfully do our best. "You have to know what to do, then do your best" is in reality a two–part challenge, a challenge for constancy of purpose and a challenge for consistency of purpose.

While constancy of purpose establishes the course for tomorrow, consistency of purpose strives to reduce the spread of that course, and helps to keep everyone's eyes on the ball. Some have compared constancy of purpose to the flow of a river where the current is constant, deep, and purposeful. The lack of consistency of purpose is likened to the river at flood stage, where the flow is out of the proper banks, spreading chillingly across territory that is foreign, slowing the flow almost to a standstill, and capable of damage.

The process of constancy of purpose is presented in the Shewhart Cycle, Figure 12–3.

PURPOSE: Start focusing on improvement of process.

POINT 14. PUT EVERYBODY TO WORK TO ACCOMPLISH THE TRANSFORMATION.

Discussion of Point 14, which Tveite has paired with "Constancy of Purpose," is presented as the summary discussion for the 14 Points for Management.

LEADERSHIP: Start focusing on improvement of process.

POINT 7. INSTITUTE LEADERSHIP. THE AIM OF SUPERVISION SHOULD BE TO HELP THE PEOPLE AND MACHINES AND GADGETS TO DO A BETTER JOB. SUPERVISION OF MANAGEMENT IS IN NEED OF OVERHAUL AS WELL AS SUPERVISION OF PRODUCTION WORKERS.

The statement of constancy of purpose for safety management is a statement of the aim of safety program management activities. The aim of leadership is to help the people, machines and gadgets to do a better job. Helping the people involves not judging or overseeing their performance, as the name supervisor implies. William Sherkenbach has commented that "In the new economic age coaching and teaching becomes the aim of supervision. The prime responsibility of the supervisor (as a leader) must be to develop his people so that they continually improve, so they can do a better job."[10]

In order to accomplish the aim of helping the people leaders must select well, train well, and educate for understanding through counseling, leading on a daily basis to create a team feeling to work toward continuous improvement. Leader must show their need for and appreciation for their help with ideas for improvements that will make jobs easier, safer, more interesting; improving the system to make it possible, on a continuing basis, for everybody to do a better job with greater satisfaction.

In his seminars, one of Dr. Deming's most frequent questions is "By what method?" By what method do leaders develop their people so that they can continually improve?

In order for the coaching and teaching to be coherent, without ambiguities, understanding of and appreciation for the system of profound knowledge is essential: appreciation for a system; knowledge about variation; the theory of knowledge; and human psychology. The system of profound knowledge is the methodology for leadership in the new economic age.

Staff safety services to operations have always been heavily involved in "safety training" of managers, supervisors, and workers. The emphasis, however, has been on technical safety information, the practical do's and don'ts provided by others, of accident and injury prevention, with emphasis on the latter. Total quality safety management, based on the system of profound knowledge, presents a new landscape for safety management activities, continual improvement of safety program and management safety systems. The aim of leadership is to develop the people so they continually improve in their new role as contributors to, and participants in, the continual improvement of these systems. In the systems of employee involvement, the team approach, leadership based on principles of human psychology within the system of profound knowledge, is the new landscape of total quality safety management. A

---

10. Sherkenback, *The Deming Route to Quality and Productivity, Road Maps and Roadblocks:* Chapter Ten.

sense of ownership and pride in the safety program will be the result. The job of leaders is to accomplish the transformation of their organizations.

LEADERSHIP: *Stop focusing on judgement of results.*

POINT 11. ELIMINATE WORK STANDARDS (QUOTAS) ON THE FACTORY FLOOR. SUBSTITUTE LEADERSHIP. ELIMINATE MANAGEMENT BY OBJECTIVE, MANAGEMENT BY THE NUMBERS, NUMERICAL GOALS. SUBSTITUTE LEADERSHIP.

The reference is to standards established for the purpose of predicting costs. Developed by industrial engineers for accounting use, such standards become the expected rate or quota on the production floor. They are usually set with an eye on the average worker, forgetful that half of the workers are above average, and half are below. Peer pressure holds the above average to the rate. The workers below average cannot make the rate. The results are a cap on production, boredom for the above average, fear for the below average, no chance for pride of workmanship, increased tendency for performance discrepancies, and disregard or circumvention of safety measures that are seen as impediments to making the quota. Deming makes the point that a quota is totally incompatible with never–ending improvement, with no trace of a system by which to help any to do a better job.[11]

Piece work and incentive pay is even more devastating than quotas, a lesson learned early in the author's safety career. Still vivid in memory is the young girl yelling to her partner "Don't slow down, keep it running, we're way ahead." as she left the department for the hospital with a severe hand and finger injury. She had been injured handfeeding a sheetmetal punch press on an incentive rate. One look at the press revealed that the guard was out of position.

The second part of Point 11 is directed to the elimination of numerical goals for people in management, and at the top of the list is elimination of management by objective. This also strikes home for the author. Again early in his career, he was one of thirteen students in the first full year of Professor Peter Drucker's career at New York University Graduate School of Business Administration. The semester year was 1952–53, one year before the publication of his landmark book, *The Practice of Management*,[12] in which the concept of MBO was introduced. So it is with mixed emotions that the author faces Dr. Deming's call for the elimination of MBO.

The problems that have prompted Dr. Deming's call lie not within Professor Drucker's concept, but in the failure of MBO practitioners to fully understand the concept. The problems that have arisen were clearly identified by Drucker in *The Practice of Management* and repeated in his later book, *Management: Tasks, Responsibilities, Practices*.[13]

While management by objectives became the accepted tag for the concept, the full title assigned by Professor Drucker is "Management by Objectives and Self–Control". Unfortunately, the latter half of the title was disregarded and its message lost in practice. But, that message was clear and unequivocal. Following is the statement on self–control as it appeared in both books. (The absence of gender neutrality is reflective of the period, early 1950's, not any bias on the part of Professor Drucker, I am sure.)

---

11. Deming, *Out of the Crisis*: 71.
12. Peter F. Drucker, *The Practice of Management* (New York: Harper & Row, 1954).
13. Peter F. Drucker, *Management: Tasks, Responsibilities, Practices* (New York: Harper & Row, 1974).

"A man's habits as a manager, his vision and his values are usually formed while performing functional and specialized work. It is essential that the functional specialist develop high standards of workmanship... . For work without high standards is dishonest; it corrupts the man himself and those around him... .

"But this striving for professional workmanship in functional and specialized work is also a danger. It tends to divert a man's vision and efforts from the goal of the business. The functional work becomes an end in itself. In far too many instances the functional manager no longer measures his performance by its contribution to the enterprise but only by his own professional criteria of workmanship. He tends to appraise his subordinates by their craftsmanship and to reward and promote them accordingly. He resents demands made on him for the sake of business performance as interfering with 'good engineering', 'smooth production', or 'hard–hitting selling.' The functional manager's legitimate desire for workmanship becomes, unless counterbalanced, a centrifugal force which tears the enterprise apart and converts it into a loose confederation of functional empires, each concerned only with it own craft, each jealously guarding its own 'secrets,' each bent on enlarging its own domain rather than on building the business.

"This danger is being greatly intensified by the technological and social changes now under way. The number of highly educated specialists working in the business enterprises is increasing tremendously and so will the level of workmanship demanded of these specialists. Our work force is increasingly becoming an 'educated' work force in which the majority make their contribution in the form of specialized knowledge. The tendency to make the craft or function an end in itself will therefore become even more marked than it is today. But at the same time the new technology will demand much closer coordination between specialists. It will demand that functional men, even at the lowest management level, see the business as a whole and understand what it requires of them. The new technology will need both the drive for excellence in workmanship and the consistent direction of managers at all levels toward the common goal."[14]

It is important to note that when Dr. Deming discusses the elimination of MBO today he employs the phrase "MBO as practiced". American managers did not heed Drucker's warning on the need for self–control. The "as practiced" version of management by objectives failed to recognize interdependencies within the organization and fostered competition, not cooperation. Dr. Deming is correct; MBO, as practiced, must be eliminated. What Drucker refers to as "self–control" and Deming refers to as "leadership" must be substituted.

Deming's call for the elimination of management by the numbers, numerical goals, is related to the elimination of "MBO as practiced". Too often, management opted for numerical objectives in the name of MBO, but with no plan for helping the people to accomplish the objective. The objective of "X percentage reduction" in the disabling injury frequency rate is a noble objective, but, as Dr. Deming would ask, "By what method?" The absence of a plan is lack of leadership and lack of self–control; it avoids the hard work that is involved in leadership and self–control.

---

14. Drucker. *The Practice of Management*.

There are, however, other problems with management by the numbers, especially financial figures. As Myron Tribus has pointed out, management by financial figures is "like trying to drive a car guided by the white line one sees in the rear view mirror. It doesn't work."[15] Worker compensation costs are, of course, a significant concern, yet they are not the data upon which to establish goals or measure performance.

And Dr. Lloyd Nelson has made the cardinal point that the most important figures for management of any organization are unknown and unknowable.[16] What gain would accrue to the organization from a safety program of which the people bragged?

LEADERSHIP: Stop focusing on judgement of results.

POINT 12(A). REMOVE BARRIERS THAT ROB HOURLY WORKERS OF THEIR RIGHT TO PRIDE OF WORKMANSHIP. THE RESPONSIBILITY OF SUPERVISORS MUST BE CHANGED FROM SHEER NUMBERS TO QUALITY.

The possibility of doing work that one can be proud of and enjoy doing is a potential that all young people hope for in life. Yet those who achieve that potential often point out that they consider themselves lucky. Why should it be considered lucky to have achieved something that should be an inherited right?

It would appear that the emotion of feeling lucky arises from the observation that so many are unlucky in this respect. Dr. Deming points out that one is born with intrinsic motivations, the desire to cooperate, a natural sense of curiosity, enjoyment in learning new things, a sense of self-esteem and dignity. These attributes are clearly visible in the very young. Yet as one grows older, forces of destruction crush these attributes to the point that those who survive them consider themselves lucky. Action to eliminate or at least minimize those destructive forces on the job is the aim of Point Number Twelve.

In Chapter 2, mention was made of the "unspoken assumption" of Frederick Taylor's theory of scientific management, "the worker's have no heads". But, the assumption went further; they also were not supposed to have any of the other characteristics that make human beings the unique creatures that they are. The industrial engineers viewed the human input as simply "hours of labor", a commodity, one of the factors of production.

In the presentation of Point Twelve in his book, *Out of the Crisis*,[17] Deming presents a litany of examples of barriers to pride of workmanship. All will be familiar to readers with knowledge of work life in a production environment. Although he does not deal specifically with barriers to pride of workmanship being operative as accident cause factors, many can be seen as relevant to the problem.

The perception the workers have of management concern about safety is directly influenced by the quality of maintenance of equipment and facilities, and especially the level of management response to reported hazards. That perception is a principle factor in determining the safety culture of the organization. In the author's experience, when the vice president of manufacturing and facilities issues and enforces a maintenance policy statement that the only items that will exceed safety-related items in priority will be mechanical breakdowns, the entire establishment responds—the

---

15. Myron Tribus, "The Quality Imperative," *The Bent of Tau Beta Pi*, Spring 1987.
16. Deming, *Out of the Crisis*: 20.
17. Deming, *Out of the Crisis*: 77–84.

maintenance crew to the feeling they are helping the people, and the people to the solid evidence that management cares.

One point of this change is one of time. The items involved will eventually be corrected. The only question is when, now or after trouble has occurred. In the meantime, they serve as a daily reminder of company policy (policy being what is done, not what is said).

A second point of the change is that injurious or damaging incidents arising from recognized but neglected hazards are not accidents arising from oversight or omissions. They are the result of risks that have been assumed by management. The correlative effect of the policy change on the safety culture of the organization is strong and immediate.

In Chapter 6, comments were directed to the matter of housekeeping as a determinant of the safety culture of the organization. The level of housekeeping will always rise or sink to the level of management's expectations as those expectations are seen by the workers. Poor maintenance practices and low standards of housekeeping are both barriers to the development of workers' pride in the organization and in their own work.

Assignment to rework, sorting good product from bad, etc., employment in the "second factory", is not conducive to pride of workmanship. How could fixing mistakes become a source of pride? It is interesting to note that whenever "the top brass" visits the plant for a tour, the rework department is noticeably empty or scrupulously avoided on the preplanned tour route. (Recall Motorola's experience with "problems hidden from management".) The "second factory" is kept in business by emphasis on the quantity produced, good product or bad.

It is also interesting to note that housekeeping receives a lot of attention just prior to such visits. Early in the author's career in a large can manufacturing plant, I questioned an assistant plant manager on the intermittent nature of this phenomena. The answer "Just to show them that we *can* do it" seemed a little off the mark to the sense of the question. The pride the employees took in having the place look nice for the visitors was clearly evident, a topic of conversation on break. My followup question, "Why not all the time?" was answered with, "Housekeeping does not make tin cans." (The author's job had been created on orders from the corporate office as the result of a very poor safety record. Turnover and absenteeism were also high.)

POINT 12(B). REMOVE BARRIERS THAT ROB PEOPLE IN MANAGEMENT AND ENGINEERING OF THEIR RIGHT TO PRIDE OF WORKMANSHIP. THIS MEANS, *INTER–ALIA*, ABOLISHMENT OF THE ANNUAL MERIT RATING AND MANAGEMENT BY OBJECTIVE.

The call for the abolishment of annual merit rating is an issue that safety managers can do little about, but for safety management this is not the point. The point being addressed is the management of people and the demoralizing effect of the practice of ranking people, teams, plants, and divisions, which is the ultimate result of merit ratings, annual reviews, and performance evaluations. Deming points out that these systems are based on the assumption that all people are alike, and he adds that the systems do not rank the performance of people, but merely the effect of the process on the people.

The idea of merit rating stirs the imagination: motivate people to do their best for their own good; get what you pay for. The result, however, is the opposite of the promise. People work for themselves. The rewards go to people who do well in the system; doing well within the system becomes the goal. Merit rating does not reward attempts to improve the system. Continual improvement of systems is inhibited by the focus on short-term results, the destruction of teamwork, the stifling of initiative or risk-taking, and the increase of variability of the performance of people, as those who are ranked below average try to better adapt themselves to the system to get a better rating. Their efforts do not improve the system, they worsen the situation. And the fact remains that in any system fifty percent will be below average and fifty percent above. Fifty percent of the people will always be in the class considered inferior. It's a system of swapping places so that all have an equal chance at being judged inferior.

Deming devotes considerable attention to the destructive force of the ranking of people, when the reality is there will always be differences between people, teams, plants, and divisions. Ranking them with rewards at the top and rejection (punishment) at the bottom is demoralizing. The problem is lack of understanding of the realities of variation.

Deming directs attention to the fact that there are intrinsic and extrinsic sources of motivation. Merit rating for bonus or recognition purposes is a form of extrinsic motivation. Rewards for designated superior or improved performances are another form of extrinsic motivation. But rewards motivate people to work for rewards. They distort the system, and foster game playing and coverups, even dishonesty. For instance, a system of annual safe driving bonuses for a large fleet of home and store delivery trucks resulted in nonpreventable decisions in 96% of accidents while liability losses skyrocketed.

Extrinsic motivation may indirectly bring positive results. Money is extrinsic reward. Some extrinsic motivation helps build self-esteem, but total submission to making money leads to destruction of the individual and the system. Ladder climbing, self-promotion to positions of power for the sake of power alone, is extrinsic motivation. It too eventually destroys.

Introduction to extrinsic motivation starts early in life; a child gets a gold star on a drawing in kindergarten, toys or money from Mommy or Daddy for doing well in school, sports, or music lessons. Children learn early to expect rewards for good performance, and as adults their desire for rewards becomes the motivating force for actions and behavior. They come to rely on things to make them feel good. Often, they later find that with all their success, their work and even their life has little meaning, and that life is not all that good. They are despondent and have low self-esteem.

The point at issue in the principle of removing barriers to pride of workmanship is using positive reinforcement to motivate people to do their best. The safe driver bonuses were supposed to motivate drivers to operate their vehicles with no preventable accidents. The result, however, was that the system made tacit liars out of drivers, supervisors, and managers as they sought to protect the bonus. In a Deming seminar, one young man rose to pose a question. "How can you expect to have a good safety record if you don't reward the people for a good record?" Dr. Deming's reply almost shook the walls. "I should be paid not to have an accident?!" There are better ways.

Those better ways will rely heavily on increasing intrinsic motivation. Intrinsic motivation arises from within, in response to inborn needs and inclinations. People are born with a need for relationships with other people, the need for love and esteem by others. They are born with a natural inclination to learn, a right to enjoyment, and a sense of dignity and self-esteem. They are also born unselfish, with a sharing attitude.

As you examine those individuals who consider themselves lucky to have found a niche in work that they are proud of and enjoy, you will find that the basis of this fulfillment rests in intrinsic motivations. If these intrinsic motivations also bring extrinsic satisfactions, so much the better, but how many people are there that are highly motivated and happy in their work with only meager extrinsic returns?

For quality safety management, these lessons need to be heeded in the design of safety program and management safety systems. An example with potential along these lines would be the system of hazard detection and correction on the shop floor. Could not these two related systems be turned over to a team of workers for redesign and development? Employees would respond positively to the opportunity to report things they believe to be hazardous to a fellow worker team member. The team, using the nominal group technique and Pareto charts, would statistically determine priorities for hazard correction. With a management safety system policy of priority rating for changing safety maintenance items or methods, timely action would result. Such a system assumes, of course, that team members would receive such education in safety and quality improvement methods needed for effective participation on the team. (Without such training, team exercises would in all likelihood have negative effects, and be little more than a farce.)

The effect would be employee ownership of that part of the safety program that most directly affects them on a daily basis. It would become a source of pride, foster a feeling of participation, and provide a sense of having some control over their own destinies on the job.

Deming's call for the removal of barriers that rob the people of pride of workmanship is a call for safety management that places a high priority on the development of program systems that foster intrinsic motivation and ferret out those elements of extrinsic motivation that result in systems of winners and losers. All must win.

LEADERSHIP: Stop focusing on judgement of results.

POINT 8. DRIVE OUT FEAR, SO THAT EVERYONE MAY WORK EFFECTIVELY FOR THE COMPANY.

Sherkenbach writes that Dr. Deming has found that the removal or reduction of fear should be one of the first of his fourteen obligations which management starts to implement, because it affects nine of his other points.[18] Following this comment, Sherkenbach presents a diagram of Deming's comment. An adapted version of his diagram is shown in Figure 12-4. In this diagram, the author, with great respect for Dr. Deming, has added one more obligation to the nine he pinpointed—Point 7, Institute Leadership.

It is highly unlikely that fear in the workplace will ever be entirely removed. The best that can be done is to make every effort to render groundless as many of the sources of fear as possible. The aim must be to remove elements that create unnecessary fears.

For quality safety management, the imperative need is for the elimination of the myth of the traditional theory of accident causation. Fear as an operational factor in safety program systems and management safety systems will not be reduced so long as the conviction remains that unsafe acts are the cause of 85% of accidents. The people need to know that they will no longer be held responsible for things over which they have no control.

The method for the transformation that is needed lies in the first two elements of the system of profound knowledge: appreciation for a system and knowledge about variation. The action that is required lies not in training but in education, education for all, starting with management at all levels.

---

18. Sherkenback, *The Deming Route to Quality and Productivity, Road Maps and Roadblocks:* 75.

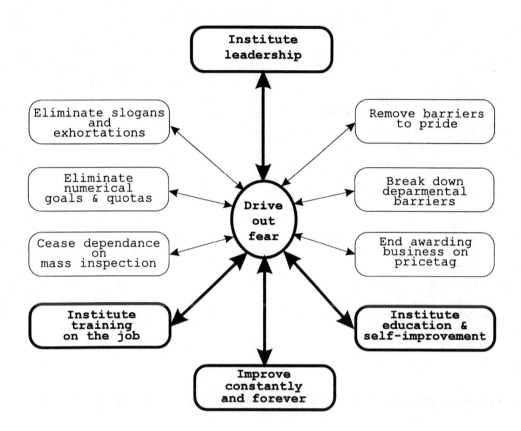

Figure 12-4. Sherkenback's diagram based on Dr. Deming's precepts for driving out fear in an organization.

The new paradigm of accident causation presented in Chapter 4 is built on appreciation for a system and knowledge about variation. It recognizes that an accident, when it occurs, is a system at work to produce harm, and that the cause of the incident, the unwanted energy transfer, resulted either from a special cause, a fleeting event that arose from outside the system, or from common causes that lie within the system, causes that are responsible for recurrent variations.

The statistical theory of system variation has established that 85% or more of system variations arise from common causes within the system. Such causes can be eliminated only by action to change the system. Actions taken to remove or influence special causes will leave the common causes undisturbed, available to cause further troubles in the future at irregular and unpredictable intervals.

Deming describes two kinds of mistakes that arise from confusing special and common causes, which can lead to disaster:

1. Attributing a variation or a mistake to a special cause, when in fact the cause belongs to the system.
2. Attributing a variation or a mistake to the system, when in fact the cause was special.

The flaw in the traditional theory of accident causation is the assumption that 85% of accidents are due to special causes, unsafe actions that are fleeting events from outside the system, with no attempt needed to determine otherwise. The result is no system improvement. The common causes remain undisturbed to continue to cause future troubles. For the workers the scenario generates not trust, but distrust of the safety system, the fear of being found at fault.

The use of fear to influence safe behavior has a long history in traditional safety programs. In the early days the approach was to graphically show the gruesome results of accidents with the message of the need for safe behavior. Over the years this approach was replaced with more subtle messages, avoiding vivid graphics but retaining the basic message that the cause of accidents and injuries is the unsafe things that people do. Such "training" films and videos remain in use today. While the approach may influence some viewers for the moment, the damaging difficulty is the workers know that the underlying message is simply not true. The message they receive is that, in the eyes of the boss, they are going to be held responsible for the results of underlying things over which they had little or no control. This is the unnecessary fear that must be driven out. It is the fears that arise out of the system that Deming refers to in Point 8, not the fear of the results.

COOPERATION: Start focusing on improvement of process.

POINT 2. ADOPT THE NEW PHILOSOPHY. WE ARE IN A NEW ECONOMIC AGE, THE AGE OF THE GLOBAL ECONOMY. WESTERN MANAGEMENT MUST AWAKEN TO THE CHALLENGE, MUST LEARN THEIR RESPONSIBILITIES AND TAKE ON LEADERSHIP FOR CHANGE.

The new philosophy is based on appreciation for a system. In organizational terms, Dr. Deming has defined a system as "a network of interdependent components that work together to try to accomplish the aim of the system." The aim of the system is contained in the statement of constancy of purpose: *to improve continually and forever the systems of production and service to become competitive to stay in business and provide jobs.* Deming points out that an aim must always relate directly to how life is better for every one. It must never be defined in terms of activities or methods.

The new philosophy is a change in focus from the management of results to improvement of processes and systems, from competition between departments and internal functions to cooperation between all functions. It is a philosophy based on achieving the economic benefits of win–win relationships. A requisite for successful implementation of the new philosophy is "system vision", appreciation for the interdependence among all functions and activities operative both within and in service to the organization or enterprise.

Safety management, the preplanned management of energy for high performance without unwanted damaging transfers, is a component of the organizational system. For quality safety management a threefold aim has been established in the statement of constancy of purpose: (1) the elimination of wastes that arise out of accidents; (2) the establishment of safety systems that nurture pride of accomplishment and self-esteem for all; and, (3) continual improvement of all safety program and safety management systems.

In quality safety management, the need is for the cooperation of all in identifying and improving the common causes of incident/accident occurrences, those causes that lie within safety program and safety management systems. People need to feel at ease and secure to make suggestions. Systems that make open discussion of problems and suggestions for improvement must be put in place. There must be teamwork. The fears that result in withholding information, covering up, and playing games must be identified and eliminated. Management must follow through on suggestions. Management alone must take responsibility for faults of the system.

The operational method for this cooperation, based on trust, is the system of profound knowledge; appreciation for system causes of accidents; knowledge about variation and its role in incident–accident occurrences; a theory of accident causation that facilitates prediction and control; and application of the principles of human psychology. Trust cannot be established in the presence of fear, nor under the circumstances of competition. Systems that encourage "willing worker" cooperation and input into the process of continual improvement of safety programs will be a source of pride of workmanship for all.

For quality safety management, the shift of focus from the management of results to continual improvement of operational safety systems requires adoption of a new understanding of "what safety is all about", a new theory of accident causation and prevention. Improved morale and increased self–esteem will result from their realization that they will no longer be blamed for things over which they had no control, and as their contributions to the improvement process become recognizable realities.

COOPERATION: Stop focusing on judgement of results.

POINT 9. BREAK DOWN BARRIERS BETWEEN DEPARTMENTS.

There are close relationships between many of Deming's Fourteen Points, themes that are repeated. One of the repeated themes is the necessity for teamwork throughout the company. Cooperation and teamwork is a major point in the discussion of fear, Point 8, and it is a major theme in the need for breaking down barriers between departments, most particularly staff areas. Dr. Deming points out that teamwork requires you to compensate for someone else's weakness with your strength, for everyone to sharpen each other's wits with questions.[19] All have much to learn from each other. All have need for the eyes of the other, and for the knowledge and experience of the other. All must communicate with each other.

One factor that works against teamwork is the tendency for departments or functional units to work for optimization of their own work and disregard, either intentionally or unintentionally, relationships that their work has on others. In large part this centered concentration is derived from the annual rating. Participation in the work of a productive team helps the company but leaves less tangible evidence to count for the individual. Who did what? Those that work to help others may not have as much to show for the annual rating as they would if they had worked alone.[20] In other cases, the centered concentration is the result of personal ego or (misplaced) pride in one's operation or area of concern. "I am in charge here. I'll run my operation, you run yours, and don't meddle with mine", a competitive, win–lose attitude.

---

19. Deming, *Out of the Crisis:* 64.
20. Ibid.

The key to breaking down barriers between departments rests in the first element of profound knowledge, appreciation for a system, including appreciation for the interdependencies of all systems. For safety management, use of the cause and effect chart and deployment flow charting will help to identify opportunities for interdepartmental cooperation, opportunities that will help break down barriers between departments as all work together to continually improve safety program and management safety systems.

As appreciation for the interdependencies of systems increases, the result will be the creation of information networks where information flows freely between departments and between teams empowered to work on and solve problems. In practice these information networks, which have their roots in the principle of breaking down barriers between departments, become integral to the entire process of continual improvement.

COOPERATION: Stop focusing on judgement of results.

POINT 4. END THE PRACTICE OF AWARDING BUSINESS ON THE BASIS OF PRICE TAG. INSTEAD, MINIMIZE TOTAL COST. MOVE TOWARD A SINGLE SUPPLIER FOR ANY ONE ITEM ON A LONG-TERM RELATIONSHIP OF LOYALTY AND TRUST.

In their discussions of Point 4, both Deming and Sherkenbach concentrate almost exclusively on how to be a good customer for the materials and supplies required for production or services such as shipping and distribution. The result is that Point 9 appears to have minimal significance for the management of safety programs and systems. Yet to the extent that safety management is involved in the direct purchase of safety equipment, supplies, and services, or serves as advisor to others in the organization so involved, the discussions have pertinence.

Deming points to the value of a long-term relationship between purchaser and supplier, a relationship built on trust and full and open communication on customer needs and supplier problems. Is there open discussion of "This is what I can do for you. Here is what you might do for me?"[21] It is his suggestion that the supplier's budgeted expense for research and development, and his past record for the development of new or improved products or services, is important. Can the supplier present evidence of management involvement in the process of never-ending improvement of processes and the other principles of transformation? Should a visit to the supplier's production facility be in order? What are the housekeeping circumstances in the production areas, warehouse areas, everywhere? Deming indicates that the Japanese strongly believe that an atmosphere of cleanliness adds to quality.

The most extensive safety program system identified in the Management Oversight Risk Tree is that of Technical Information, and the largest collective expenditures for many safety management programs is for technical information in one form or another. On a personal professional basis this includes society and association memberships; attendance at conferences, symposiums, and seminars; enrollment in university or extension courses; and safety periodicals and journals. On an impersonal basis are technical information sources such as information services related to regulatory requirements, reference books and manuals, consultant services, and so on. For quality safety management, the difficult question is "How does one become a good customer for these technical information services?"

---

21. Deming, *Out of the Crisis*: 43.

The answer for the personal information sources lies in becoming an active participant in their processes. A definitive answer for the passive sources of technical information is more difficult. Both situations may seem unrelated to Deming's call for ending the practice of rewarding business on the basis of price tag, but this is his red flag to call attention to the need to be a quality minded customer.

TRAINING & EDUCATION: Start focusing on improvement of process.

POINT 6. INSTITUTE TRAINING ON THE JOB.

POINT 13. INSTITUTE A VIGOROUS PROGRAM OF EDUCATION AND SELF-IMPROVEMENT.

Training and education is the heart of the transformation. Training and education pump the blood that brings life to and keeps alive the program of transformation. They are the essence of the aim to change management methods completely or essentially in composition or structure. This is so because the transformation is totally based on changes that occur in the mind.

Training and education are intimately interrelated. Webster's unabridged yields the following pertinent definition of training: "the development of a particular skill or group of skills, instruction in an art, profession or occupation."[22] Deming notes that Point 6 refers to the foundations of training for management and for new employees.

Education, on the other hand, has a broader impact. According to Webster, to educate is "to develop by fostering to varying degrees the growth or expansion of knowledge, wisdom, desirable qualities of mind or character, physical health or general competence, especially by means of a formal education." Deming notes that Point 13 refers to continual education and improvement of everyone on the job—self-improvement.

Comparison of the definitions indicates that training is certainly an element of education, but training does not have the defining objective of expanded knowledge or wisdom, although these attributes will accrue to many training activities. The defining objective of training is immediate need. The defining objective of education can be stated as preparation for future needs.

Both principles, the call to institute training, and the call for education and self-development, are calls for the continual development of the most valuable asset in the organization, the people. And, Dr. Deming makes it clear that the calls are intended to include all of the people, from executive and general management to line management and staff to the workers, each at his or her appropriate level of competence and expectation.

For the transformation, the bedrock foundation for both training and education is the system of profound knowledge:

- Appreciation for a system
- Knowledge about variation
- Theory of knowledge
- Human psychology

For quality safety management, the aim of this book has been directed to these cardinal points of the system of profound knowledge: appreciation for the system of incident/accident occurrences, and appreciation for safety program and management safety systems; knowledge about variation and the role of common causes and special

---
22. *Webster's Third New International Dictionary of the English Language* (1986).

causes in the hazard related incident/accident system; a theory of incident/accident causation that enables prediction and control; and the application of human psychology in employee involvement and team methods of problem identification and problem solving.

Just as the call for the abolishment of annual merit rating is an issue that safety managers can do little about, so too is the call for training and education from the executive level to the workers. However the call for a program of self–improvement is well within the province of individual action.

Most safety professionals probably regard membership and participation in the educational activities of such organizations as the American Society of Safety Engineers, National Fire Protection Association, National Safety Council, American Industrial Hygiene Association, and others a program of self–improvement. At this point, however, the difference between training and education becomes an issue. Is participation in these activities a response to current need, or for the purpose of expanded knowledge and wisdom in preparation for future needs?

Future needs are already strongly in evidence. Reference has already been made to Peter Drucker's comment that the widespread involvement of the people in the continuous improvement of work is changing the social organization of the factory (page 161). Drucker expands on this development with the following comments.[23]

> "The Japanese employ proportionately more machine operators in direct production work than Ford or GM. In fact, the introduction of SQC almost always increases the number of machine operators. But this increase is offset many times over by the sharp drop in the number of non–operators: inspectors, above all, but also the people who do not do but fix, like repair crews and 'fire fighters' of all kinds.

> "Moreover, the first–line supervisors also are gradually eliminated, with only a handful of trainers taking their place. In other words not only does SQC make it possible for machine operators to be in control of their work, it makes such control mandatory. No one else has the hands–on knowledge needed to act effectively on the information that SQC feeds back."

Evidence of the reality of Professor Drucker's comment appears with regularity in the business press, such as *Business Week, Fortune, Forbes,* etc., and in the business pages of daily newspapers, such as the excerpts from *The St. Louis Post Dispatch*[24] on the next page.

A program of self–improvement based on additional training in safety and other related occupational techniques cannot be regarded as a program of preparation for future needs. It is a maintenance program, designed to remain current with the state of the art. It is not a program of growth or expansion of knowledge or wisdom.

A program of self–improvement based on education, "the growth or expansion of knowledge, wisdom, desirable qualities of mind or character, physical health, or general competence," will be based on learning from information that comes from outside disciplines or sources. This is the significance of the phrase "especially by means of a formal education" that concludes the Webster definition. For total quality safety management this is the significance of Deming's system of profound knowledge. The four elements of profound knowledge are the outside sources of the knowledge required for continual improvement in the management of the systems for the elimi-

---

23. Peter F. Drucker, *Managing for the Future* (New York: Truman Talley Books—Dutton, 1992).
24. Christopher Cary, *The St. Louis Post Dispatch,* September 24, 1993.

> ## A–B Charts New Course for Business
> ## Teamwork a Key Element*
>
> Anheuser–Busch Cos. Inc. has settled on technology and team work as its answers to weak demand and increasingly tough competition in the domestic beer market.
>
> The organizational and operational changes A–B announced Wednesday, along with the elimination of 1200 salaried jobs are designed to boost profits by increasing efficiency... .
>
> The company...plans to modernize its breweries and introduce team–oriented work methods to improve quality and productivity and increase employee involvement.
>
> A–B intends to empower its workers by training them to fill multiple roles and allowing them to operate with less supervision, said Jerry E. Ritter, executive vice president and chief financial officer.
>
> "People will have responsibility for their jobs, and authority." he said.
>
> A–B apparently is pleased with the results of initial efforts at its newest brewery. "The brewery, which began production earlier this year, has the type of management style and employee involvement that the company would like to see at other operations," said William L. Rammes, vice president for human resources.
>
> The new approach has been a hit with brewery workers. "They opened with 40 percent less plant supervision and gave workers more responsibility." said Bud Benack, a trustee for Teamsters Local 1129, which represents all 300 production employees.
>
> If the approach succeeds in a new brewery with new workers it should also succeed in an old brewery with veteran workers, Benack said.
>
> "They don't need supervisors to tell people who have been doing their job for 25 years what to do." he said.
>
> The 1200 jobs A–B intends to eliminate, through early retirement and attrition, are salaried positions.

*St. Louis Post Dispatch, September 24, 1993.

nation of wastes of all types that arise out of unwanted energy transfers. A program of self–improvement based on education is the requisite preparation for future needs.

IMPROVEMENT OF PROCESS: Start focusing on improvement of process

POINT 5. IMPROVE CONSTANTLY AND FOREVER THE SYSTEM OF PRODUCTION AND SERVICE.
In the late 1970s, William E. Conway, then President and CEO of Nashua Corporation, was the first American top executive to embrace and implement the Deming Method of Management. His story was given prominence in the 1980 landmark NBC telecast "If Japan Can, Why Can't We?" In 1981, at a conference with executives from Ford Motor Company, Bill Conway was asked where one would start the process of focus-

ing on continual improvement. Conway immediately referred to the Fourteen Points and, in a videotape of the meeting, appears to randomly select Point 8, Drive Out Fear, as a potential place to start.

Dr. Deming, in the workbook for his four day seminar, "Quality, Productivity and Competitive Position," struck the same note with a different expression. His statement was directed to the overall prerequisite for the management of quality and productivity: create an environment in which everybody may take joy in their work.[25] Conway's choice of Point 8 as the place to start may not have been all that random.

Figure 12-4 provides further evidence of the importance of Point 8. The central point that each was stressing, without saying so, was the need to remove the barriers to pride and foster intrinsic motivation. If it is an objective for total quality safety management to create a safety program which employees are proud of and boast about, Point 8, drive out fear, and the creation of a work environment in which everybody may take joy in work is the place to start to improve constantly and forever the safety program systems.

Following his suggestion to start with Point 8, Conway pointed to the Shewhart Cycle as the method for initiating the improvement. Sherkenbach has observed that the Shewhart Cycle is really a spiral, a spiral that rises ever upward to higher and higher levels of improved performance.[26]

In discussing the need to drive out fear, the point was made that for total quality safety management the first imperative is the elimination of the myth that 85% of accidents are caused by the actions of people. Step One of the Shewhart cycle is "Development of a Plan" for the change. For the objective of improvement of safety program systems, this will require identification of a system, or systems, where the improvement will impact on that myth, and initiate implementation of "system vision" and appreciation of system variation.

The need for broad input into the process of selecting the most suitable safety program system to initiate the program dictates that the process be a team project. Construction of a deployment flow chart will point the way at this stage. As discussions are initiated, the Safety Program System (SPS) flow charts in Chapter Nine will guide evaluation for potential impact on the objective, and final selection can be made using the nominal group technique and scaling by Pareto charts as discussed in Chapter 10.

One prime system candidate would be the "Did Not Detect/Correct Hazards" system under Operations Management and highlighted in chart SPS-3. Assuming the Hazard Detection/Correction system is selected for the project, SPS-3 and the related User Manual questions in Chapter Seven, p. 61 ff, will guide team deliberations and development of a plan of action. At this point it will be wise to remember that the MORT diagrams and *User's Manual* predate widespread use of teams in problem identification and solution. It is the sense of the ideas and the logic path that is important, not literal translation. These are guides to action, not a prescription for cure.

Once a plan is developed, and the changes introduced for trial, the Shewhart Cycle steps of Study and Action (adopt the change, abandon the change, or adjust the change for further testing) will follow. Restudy, Evaluation and Adjustment, Steps 5 and 6, will create the spiral and yield continual improvement. The ultimate objective is a system of hazard detection/correction that all are proud of and boast about, and a system that continually improves. As the people develop a sense of ownership of the safety program compliance problems will disappear as the result of peer pressure.

---

25. Author's notes from Dr. Deming's seminar referring to Section 16, Point 5 in *Out of the Crisis*.
26. Sherkenback, *The Deming Route to Quality and Productivity, Road Maps and Roadblocks*: 35.

IMPROVEMENT OF PROCESS: Stop focusing on judgement of results.

POINT 3. CEASE DEPENDENCE ON MASS INSPECTION TO ACHIEVE QUALITY. ELIMINATE THE NEED FOR INSPECTION ON A MASS BASIS BY BUILDING QUALITY INTO THE PRODUCT IN THE FIRST PLACE.

The frame of reference of Point 3, as stated, is that of quality of product. The objective of safety management is not a product, it is a service. Does Point 3 have relevance for total quality safety management? Perhaps rewording the principle will help:

CEASE DEPENDENCE ON INSPECTION TO IMPROVE QUALITY. ELIMINATE THE NEED FOR INSPECTION BY BUILDING QUALITY INTO THE SERVICE IN THE FIRST PLACE.

The purpose of safety inspections is hazard detection/prevention. Some inspections are mandated by law or regulation; for elevators, hoisting equipment, boilers and types of machinery, electrical switchgear and power transmission apparatus, pressure vessels, reactors, etc. The characteristic common to all of these things is very high levels of energy, where an unwanted transfer could be catastrophic. Another characteristic is that the detection of hazardous conditions or developing conditions that have the potential to become hazardous requires varying degrees of expert knowledge. There will always be these inspections, and they are not subject to the principle of ceasing dependence on inspection.

Here, attention is directed to routine safety inspections of the workplace to detect exposures to unwanted transfers at lower energy levels. The traditional safety inspections by the workplace supervisor, the location safety coordinator, or a safety inspection committee of one kind or another are the focus.

In the preceding discussion of Point 5, it was suggested that the safety system of Hazard Detection/Correction might be a good candidate system for initiating the transformation. Routine safety inspections of this type are primarily a device to periodically concentrate attention on hazardous conditions or practices that have been slipping into existence unnoticed. In addition to the benefits of detecting such hazards, the inspection procedure itself serves to visibly demonstrate management concern for safety, but not only to the workers. The written reports also create evidence of that concern for regulatory authorities.

However, few of the hazards so identified actually slip into the system unnoticed. The workers know about them. The workers not only know about practically everything the inspectors identify, they know also know about a lot of things that the inspectors miss. They know about them because they see them every day, and often they are conditions they are forced to work around. This includes the circumstances of poor housekeeping.

For total quality safety management, the call for ceasing dependence on safety inspections for periodic identification of workplace hazards can be met most productively by new systems which depend on educated observational reporting by everyone on a continuing basis. In other words, put everyone on the inspection team. And, following the team approach, let the reporting be done to an employee hazard control team, a team with responsibility for scaling the observations and establishing action priorities. Management has the responsibility to enforce the priority selections.

For management it matters little what is done first, what does matter is prompt and consistent enforcement of corrective action.

The primary responsibility of total quality safety management in such a system is revealed in the phrase "educated observational reporting". This responsibility is covered in Point 6, Institute Training on the Job. Specific technical content for this training for employees is discussed in Chapter 6. However, this content should be relayed within the context of appreciation for a system, knowledge of variation, and knowledge of the system theory of accident causation. Employees need to receive training in these areas at a level consistent with their competence and expectations (Point 6). Employee fear of being held responsible for things over which they have no control must be eliminated (Point 8). Employee pride in continual improvement of safety conditions and the safety program must be nurtured (Point 12).

IMPROVEMENT OF PROCESS: Stop focusing on judgement of results.

POINT 10. ELIMINATE SLOGANS, EXHORTATIONS, AND TARGETS FOR THE WORKFORCE ASKING FOR ZERO DEFECTS AND NEW LEVELS OF PRODUCTIVITY. SUCH EXHORTATIONS ONLY CREATE ADVERSARIAL RELATIONSHIPS, AS THE BULK OF THE CAUSES OF LOW QUALITY AND PRODUCTIVITY BELONG TO THE SYSTEM AND THUS LIE BEYOND THE POWER OF THE WORKFORCE.

"Safety First" is probably the most widely recognized safety slogan ever, created by a large steel company around the turn of the century. However, it is not the first safety exhortation ever created. Sherkenbach relates that the oldest he has ever seen was inscribed in an extremely rickety staircase in a 13th century castle on the Rhine river. Translated, it reads:

> O God, I beg
> Guide my step
> So I do not fall.[27]

Safety posters and exhortations have long been a favorite element in organized safety programs. In fact, their production and promotion is almost an industry in itself even to this day. The *Accident Prevention Manual for Industrial Operations*, discussing the purposes of safety posters presents the following:[28]

> 1. To remind employees of common human traits that cause accidents;
> 2. To impress people with the good sense of working safely;
> 3. To suggest behavior patterns that help prevent accidents;
> 4. To inspire friendly interest in the company's safety efforts;
> 5. To foster the attitude that accidents are mistakes and safety is a mark of skill;
> 6. To remind employees of specific hazards.

Deming's statement is that "posters and exhortations generate frustration and resentment. They advertise to the production workers that management is unaware of the barriers to pride of workmanship."[29] The purposes of the types of posters presented above do more than that; they instill and reinforce the myth that the predominant cause of accidents are the things that people do. They stem from management's conviction that workers could, by working more safely, prevent almost all accidents.

---

27. Sherkenback, *The Deming Route to Quality and Productivity, Road Maps and Roadblocks*: 83.
28. *Accident Prevention Manual for Industrial Operations, Administration and Programs*, Ninth Edition (Chicago: National Safety Council, 1988).
29. Deming, *Out of the Crisis:*. 67.

A latter day exhortation that has entered management's vocabulary is the objective of zero defects. This has been adopted by safety as a call for error–free performance.

While a campaign of posters, exhortations and pledges to work more safely may result in some fleeting improvement and the elimination of some obvious special causes, no permanent improvement will occur in the absence of improvement in the safety program and management safety systems. System improvement is the responsibility of management. Entreating the workers as a method for improvement cannot result in continual improvement. The workers will see the effort as being asked to do what they are unable to do. The effect is fear and mistrust of management.

Dr. Deming makes the point that goals that an individual may set for him or herself are not only helpful but necessary for accomplishment. They give direction and meaning to one's efforts. The goal of a college education will create resolve to study harder for good entry marks, or to qualify for merit scholarship. But goals set for other people, without a method for achieving the goal, have effects opposite to those sought. "By what method?" is the Deming question. To have meaning, a goal must include a method of accomplishment that is within the capability of the intended achiever.

Posters that would explain to everyone what is being done to improve safety program and management safety systems would be another matter. Posters that would update the people on the results of team efforts for safety program improvements would be meaningful: improved systems of hazard reporting, more effective response on hazard elimination, improved layouts, ergonomic and environmental redesigning, improved task instructions, etc. People would then understand that management is taking responsibility for hang–ups, defects, procedures that enhance chance taking, and other obstacles that affect their safety.

Posters can be effective communicators if the right message is directed to the right people. Dr. Deming says, "If you want to have posters, let the people decide what posters to put up."

PURPOSE: Start focusing on improvement of process.

POINT 14. PUT EVERYBODY TO WORK TO ACCOMPLISH THE TRANSFORMATION. THE TRANSFORMATION IS EVERYBODY'S JOB.

Now comes the hard part. It is difficult because there is no organizing theory on accomplishing the transformation, not even on getting started. It is probable that no two organizations have initiated their programs in the same way, and no more probable that two identical programs have developed independently of each other.

However, two observations can be offered. The first is that the transformation starts with the individual. And the second is that the transformation is based on knowledge that comes from outside, by invitation. No system can understand itself. Transformation will not be spontaneous; the process must be managed.

The key to the transformation is the system of profound knowledge. The system of profound knowledge is the method for achieving optimization, continual improvement of the operating systems that make up the enterprise. For total quality safety management, this involves continual improvement of both the safety program systems and the management safety systems. The system of profound knowledge is the new environment for total quality safety management.

The system of profound knowledge is the new environment for total quality safety management. Management of any system, including safety program and management safety systems, requires "system vision" and recognition that variation in all systems is normal.

Activities to control undesirable system variations require management, and management in any form is prediction. To be successful, prediction must be based on knowledge and understanding derived from organizing theory. In the absence of theory all that exists is information. The successful management systems that involve people require understanding and employment of human psychology.

The fourteen points are presented by Dr. Deming as principles for transformation that follow naturally as application of the system of profound knowledge. The first step is transformation of the individual. (Reference Point 13, Institute a vigorous program of education and self-improvement.)

In his book, *The New Economics for Industry, Government, Education*, (p. 95) Deming points out that:

"Once the individual understands the system of profound knowledge, he will apply its principles in every kind of relationship with other people. He will have a basis for judgement of his own decisions and for transformation of the organization that he belongs to. The individual, once transformed, will:

- Set an example.
- Be a good listener, but will not compromise.
- Continually teach other people.
- Help people to pull away from their current practice and beliefs and move into the new philosophy without a feeling of guilt about the past."

In offering these comments, Dr. Deming is addressing the subject of leadership (reference Point 7, "institute leadership"). The job of a leader is to accomplish transformation in the organization. The leader understands why the transformation will result in benefits, not only to the organization but to all of the people in the organization that he or she deals with. This results in a feeling of obligation to accomplish the transformation, an obligation both to yourself and to the organization.

And, finally, the leader is a practical person with a plan, step by step. In presentation the plan must include both a method for accomplishment and predictions of measurable results, results that will satisfy identified customer needs. The leader also knows that the best plans are those that are simple and brief in presentation. Small bites at a time are more easily digested than gorging on an overloaded platter. In short, the leader must manage the transformation.

The Staff Function Paradox[30]

For safety management, does not the edict "put everybody to work to accomplish the transformation" present a paradox? In discussing the role of a manager of people, Dr. Deming identifies three sources of power that a manager possesses:

1. Authority of office
2. Knowledge
3. Personality and persuasive power; tact.[31]

---

30. "The Staff Function Paradox," is the title of an article by William K. Fitzgerald, a manager of organization development at Hewlett Packard's Little Falls site, published in *The Journal for Quality and Participation*, vol 16, no. 3, June 1993, published by AQP—Association for Quality and Participation, 801–B West 8th Street, Cincinnati, Ohio 45203. His thoughts were most helpful in addressing this subject.

31. Deming, *The New Economics for Industry, Government, Education*: 129.

While his statement accurately reflects the conditions for the line manager, it does not hold for the staff manager. Staff managers have no ordained authority of office. Safety managers, in electing to serve in this functional specialty, have in effect traded the need for power and control to do interesting, exciting work.

Once in service, those who serve in functional staff capacities, including safety managers, become keenly aware of the ramifications of this decision. One common result of the lack of ordained authority of office that can be observed is the establishment of ersatz authority through policies and procedures. These policies and procedures are the product of the specialized knowledge that the staff functionaries possess. The result can be the creation of kingdoms unto themselves as power is wielded through policies and procedures.

The call for putting everyone to work to accomplish the transformation is a call for greatly increased levels of involvement. For safety management, it is a call to increase the level of safety expertise in areas thought to be the sole domain of the safety function. This requires a rethinking of the key role of safety managers as inside consultants: to work themselves out of a job. The call is to work in such a way that what the safety manager knows is transferred to the organization. The objective is to render the organization more capable of managing itself without being dependent on a consultant.

In quality safety management, the safety manager does not "put people to work". The role of the quality safety manager is enablement of the people, people at all levels, to preplan the management of energy for high performance without unwanted, damaging transfers, and to maintain constancy of purpose in the elimination of wastes that arise out of hazard related incidents; in the establishment of safety systems that nurture pride of accomplishment and self–esteem for all; and in continual improvement of all safety program and management safety systems. In this role the total quality safety manager will serve as consultant, teacher, change agent, and confidant. He or she must be a business person first and a specialty person second. Failing the role of enabler, the only activity that must be filled is keeping the enterprise on the right side of the law.

Index

–A–

Accident
    amelioration LTA, 78–82, 116
    anatomy of an, 34–35
    causation, theory of, 15
    incident system LTA, 58
    need for a definition of, 25–26
    prevention strategies, 107
    prevention of second, LTA, 78–79
Accident Prevention Manual for Industrial Operations, 13, 157–58
Accident Triad, 30–32, 45–49, 148
    MORT analytical logic of the, 45–49
Accident/injury rated investigation procedures, 136–39
Accountability LTA, 90
Amelioration LTA (post–accident), 116
American Industrial Hygiene Association, 188
American Society of Safety Engineers, 188
Anheuser–Busch, 189
Assumed Risk (AR), 47
    MORT legend for, 136
    oversights and omissions vs., 101–102
Audit(s), 100
    independent, and appraisal LTA, 60
    upstream process, LTA, 59

–B–

Bar chart, 153
Barrier(s)
    concepts of energy and, 37–51
    less than adequate, 137
    and safety program systems, 35, 48
    tactics of results management by, 42
    on the target, 46
Bell Telephone Laboratories, 4
Bergstein, Melvyn E., 167
Bird, Jr., Frank, 166

Block function and work schematics LTA, 101
Blue Ribbon Team (critical and catastrophe rated), 138–39
The Bridge, 82–84, 117, 130, 135–36
Brodeur, Paul, 12
Budgets LTA, 90
Business Week, 167–68

–C–

Carnegie Mellon University, 167
Cause and Effect Diagram (Fishbone Diagram), 142–47, 150
    PEEMM example of a, 144
Checklists, 65
Classified accident/injury investigation, 132–35
Common causes, 151
Communication LTA, 57, 82
Compulsory Worker's Compensation, 12
Concepts and requirements LTA, 119
Contributory Negligence, 10
Control(s)
    charts, 151–53
    energy, procedures LTA, 94
    safety analysis, 72–73
Conway, William E., 189–90
Cooperation between departments, 184–87
Coordination LTA, 84
Crawford–Mason, Clare, 5
Culture of good housekeeping, 44

–D–

Data
    collection and analysis LTA, 59
    controlling the quality of, 152
Delays, 90
Deming, Dr. W. Edwards, 1, 4, 19, 82

call for the elimination of numerical
  goals by, 178
classification of special causes by, 22
Fourteen Points for Management,
  163, 171–94
instructions to flow diagrams, 155
on leadership, 194
ratio of system variation, 151–52
role of intrinsic and extrinsic
  motivation recognized by, 181–82
statement on posters and
  exhortations, 192
System of Profound Knowledge,
  169–73, 188–89
theory of management, 169
three sources of management power,
  194
training and education stressed by,
  187
Deming Cycle, 4
*The Deming Route to Quality and
  Productivity, Road Maps and Roadblocks,*
  172
Departmental team (serious rated),
  137–39
Deployment Flow Charting, 141–42, 150
Design
  arrangement LTA, 97
  and development plan LTA, 94–100
  process LTA, 98–99
Detection of hazards, 64–66
Deviation(s)
  causes of, 22
  consideration of, LTA, 75
Diagnostic statistical analyses LTA, 59
Directives LTA, 89
Discrepancy (error) sampling system
  LTA, 58
Dobyns, Lloyd, 5
Drucker, Peter, 159–61, 188
  concept of management by
    objectives and self–control, 177–78

–E–

Education and training, 187–89
Emergency
  action LTA, 78–79
  provisions LTA, 98

Employee(s). *See also:* Management;
  Process improvement; *and* Safety
  program
    cooperation, 184–87
    fear, 182–84
    involvement in safety program
      improvements, 157–68, 169–95
    motivation LTA, 75–78
    pride of workmanship, 180–82
    role of, in TQM transformation,
      193–95
    training and education, 187–89
    work quotas, elimination of, 177–79
Energy
  and barriers, concepts of, 37–51
  controlling, 42–43, 94–95
  interference with normal types of,
    39–40
  types and sources of, 38–40
Energy flows
  functional and nonfunctional, 47
  non/precedents for the prevention of
    unwanted, 56
Environment LTA, 97
Errors, failure to predict, 96
Execution LTA, 63
External communication LTA, 57

–F–

Fear, as an impediment to working
  effectively, 182–84
Fellow Servant Rule, 10
First Aid LTA, 80
Fishbone Diagram, 142–45
Fisher, R.A., 19
Florida, Richard L., 167
Flow chart, 127, 148–49
Ford Motor Company, 189
The Fourteen Points for Management
  (Deming), 163, 171–94

–G–

Galbreath, Frank and Lillian, 14
General Electric, 167
Germany, development of social security
  laws in, 11
Gibson, James J., 40
Goals LTA, 91
Guide to Accident/Injury Rating, 132–33
Guide to Quality Control, 144

–H–

Haddon, Jr., William, 40–41, 164
HAP. *See:* Hazard Analysis Process (HAP)
Harley–Davidson, 165
Hazard(s)
    analysis process triggers LTA, 60
    correction of, 66
    detection/correction, 64–66, 110, 191
Hazard Analysis Process (HAP), 60, 131
    concepts and requirements of, LTA, 91–100, 119
    HAZRIN(s), 46–47, 128
    defined, 31
    prevention of the, 53
    wastes that accrue to, 174–75
Health, monitoring general, LTA, 59
Heinrich, H. Waldo, 14–15, 17, 19, 24
Higher supervision services LTA, 83, 89–90
Histogram, 153–54
Housekeeping
    culture of good, 44
    program LTA, 66
Human factors review LTA, 95–96

–I–

Incident(s)
    investigation, 136–39
    services, 137–39
Incident/accident investigation(s), 128–30
    departmental team, 137–39
Information
    exchange LTA, 84
    flow LTA, 89
    research LTA, 94
    systems LTA, 91, 108
Injury
    potential by energy level, 134
    severity, 133
Inspection
    LTA, 96
    mass, dependence on, 191–92
Interdepartmental coordination LTA, 66
Internal communication LTA, 57
Investigation procedures, accident/injury rated, 136–39
Investigation team(s), 135–39, 140–42
Ishikawa, Kaoru, 142, 144, 147
Ishikawa Diagram, 142

–J–

Japan, Quality Circles in, 158
The Japanese System, 1
Job Safety Analysis (JSA), 68. *See also:* Task Safety Analysis
Johnson, William G., 27, 40, 130
Joiner, Brian, 158
JSA. *See:* Job Safety Analysis (JSA)
Judgement of results, 177–80
Juran, Joseph M., 19, 82, 106
    application of the Pareto Principle by, 131, 137–8
    classification of problems by, 137–38
    discussion of the Trivial Many by, 140

–K–

Key personnel injury, 133–34
Knowledge, theory of, 170
Kuhlman, Raymond L., 132

–L–

Leadership, 177, 179, 182, 194
    in focusing on improvement of process, 176
Less Than Adequate (LTA), 55
Line responsibility LTA, 89
Logs LTA, 65
LTA. *See:* Less Than Adequate (LTA)

–M–

Maintenance/inspection LTA, 62–63, 96, 109
Management. *See also:* Total Quality Management
    cooperation, 184–86
    coordination LTA, 84
    economic imperative of quality, 165–66
    by fact, 127
    focus on judgement of results, 177–82
    leadership initiatives under Deming's Fourteen Points for Management, 176–85
    by objectives, 177–82
    operations, LTA, 63–78, 110
    peer committees LTA, 101
    power, sources of, 194–95

responsibilities of executive, 87–103
of risk, 13–14
safety responsibilities of executive, 85, 87–103
service, SPS–9, 136
supervision services LTA, 83–85, 89–90
team responsibilities of, 135–36
theory of scientific, 14
Management: Tasks, Responsibilities, Practices, 177
Management Oversight Risk Tree (MORT), 27–30, 127–28, 140
   analytic logic of the Accident Triad, 45–49
   concern, vigor and example LTA, 76
   description of wastes in the, 50
   graphic symbols and legends used in the, 55
   HAP triggers postulated by the, 60
   legend for assumed risk, 136
   performance errors, 67
   restated, 54, 105–126
   safety training recognized in the, 163
   scaling devices used with the, 131–32
   statistics used in the analytic logic of the, 160–62
   summary, 103
   technical information identified in the, 186
   top events, 28–30
Management safety systems, 118–25, 130
   implementation of, LTA, 118
   LTA, 87–88, 118
Manuele, Fred, 30–31
Man–machine requirements, 96
Material type, 134
McFarland, Ross, 37–38
Measure of performance LTA, 84
Medical services LTA, 80–81
Merit rating, 180–81, 188
Monitor points LTA, 99
Monitoring
   audit and comparison LTA, 100
   systems LTA, 57
MORT. *See:* Management Oversight Risk Tree (MORT)
MORT User's Manual, 55, 190
Motivation
   general, program LTA, 77
   intrinsic and extrinsic, 181–82

Motorola, 154, 159–60, 166–67, 180

–N–

National Fire Protection Association, 188
National Private Truck Council, 160
National Safety Council, 13, 25, 188
   Accident Prevention Manual for Industrial Operations, 13, 157–58
Nelson, Lloyd, 179
The New Economics for Industry, Government, Education, 194
New York University Graduate School of Business Administration, 177
Nominal Group Technique (NGT), 145
Nonserious incidents, 136–37
Normal variation, 23
Numerical goals, elimination of, 177–79

–O–

Occupational Safety and Health Act, 158, 162
Operations management LTA, 110–115
Operational readiness LTA, 61–62, 109
Operational specifications LTA, 97–98
Out of the Crisis, 172, 179
Oversights and omissions vs. assumed risks, 101–102

–P–

Paradigm shift, 1–7
PDSA Cycle, 4
Pareto
   charts, 141, 147–48
   Distribution, 139
   Principle, 131–32, 138, 140, 166
Personnel
   motivation LTA, 99
   organization of safety, 100–101
   qualification and training LTA, 99
   responsibilities of, 89
   selection LTA, 97
Personnel performance discrepancies (errors), 67–78, 73–78, 96, 111–15
Point of operation log, 63
Pope, William C., 25–26
Posters, 192–93
Potentially Harmful Incident, 46–47. *See also:* HAZRIN.
The Practice of Management, 177

Pre–task briefing LTA, 73
Principle of the Vital Few, 131
Principles of System Variation, 151
Priority problem list LTA, 59
Process control chart, 23
Process improvement, 176–77, 184, 169–94. *See also:* The Fourteen Points for Management; System of profound knowledge; *and* Variation
    cease dependence on inspection for, 191–92
    constant, 189–91
    cooperation necessary to achieve, 184–87
    creating constancy of purpose required to achieve, 174–76
    training and education necessary to achieve, 187–89
Professional Accident Investigation, 132
Professional staff LTA, 101
Program aids LTA, 84
Psychology, role of human, in Deming's management theory, 170–71

–Q–

Quality circles, 158

–R–

Rating incidents, 137–39
Readiness, verification of use, LTA, 61–62
Red Ribbon Team (severe rated), 137–38
Rehabilitation LTA, 81
Reiner, Carl, 167
Reported Significant Observation (RSO) systems, 58, 167
Rescue LTA, 78–80
Research and fact finding LTA, 84
Resources LTA, 84–85
Risk
    assessment system LTA, 90–102, 119–25
    assumption of, 10
    bearing, 12–15
    management, 13–14
    protection analysis LTA, 59
    response LTA, 85
Roosevelt, Theodore, 10
Rule of System Variation, 151
Run chart, 20–21, 151
Runaway Truck Accident, 129–30

–S–

Safety
    analysis criteria LTA, 91–93
    committee, 157–58
    controls, 72–73
    engineering, foundation of, 14
    folklore of, 15–17
    inspections, 58, 191–92
    management, 33–35, 174, 195
    observer plan LTA, 58
    organization LTA, 101
    policy statement, 88, 100
    posters, 192–93
    procedures LTA, 93
    requirements LTA, 93–94
    responsibilities of executive management, 85
    task, LTA, 68–73
    training, 163–64, 176
Safety management
    professionals, statement of constancy of purpose for, 174
    role of, 195
    tactics and strategies of, 33–35
Safety program, 100–101
    activities, 49–51
    implementation LTA, 88–90
    improvements, employee involvement in, 157–68
    organization for improvement LTA, 100–101
Safety program systems, 107–117, 130, 190. *See also:* Monitoring systems *and* Total Quality Management
    and barriers, 35, 48
    diagram of basic, 53
    methodology for continual improvement of, 168
    performance discrepancies (errors) in, 111–15
    strategies of prevention by, 53–85
St. Louis Post Dispatch, 188–89
Scaling devices, need for, 131–32
Scatter (Correlation) diagram, 154–56
Schematics LTA, 65, 100
Scholtes, Peter R., 9, 160
Serious incidents, 137–39
Sherkenback, William, 172, 182–83, 186
Shewhart, Walter, 19, 20–22, 24

Principles of System Variation, 151–52
Shewhart Cycle, 3–6, 175–76, 190
Sholtes, Peter R., 9
Slogans, elimination of, 192
Smith, William B., 159
Special causes, 22, 151–52
    confusion of, with common causes, 183–84
The Staff Function Paradox, 194–95
Standards and directives LTA, 84
Statement of Constancy of Purpose for Safety Management Professionals, 174
Statistics, 20–24
Statistical analyses, diagnostic, LTA, 59
Statistical methods,
    key, for specific thinking, 147–50
    overview of analytical, 150–56
    used in the analytic logic of the MORT, 160
Statistical quality control (SQC), 161, 188
Strategies of prevention by safety program systems, 53–85
Supervision
    LTA, 97
    services LTA, 83, 89–90
Supervisor
    judgment LTA, 66
    monitor plan LTA, 65
    transfer plan LTA, 64
Suppliers, building long-term relationships with, 186–87
System(s)
    appreciation for a, 169
    monitoring, LTA, 57–58
    observer plan LTA, 58
    vision, 162–63, 190
System of profound knowledge (Deming), 169–71, 182–84, 193–94
    education and training in the, 187

–T–

Task(s), 96. *See also:* Task safety analysis
    assignment LTA, 67
    performance discrepancies, 67–68, 111–15
    procedures did not meet criteria, 98–99
    schedules, 76
Task safety analysis program LTA, 68–73

Taylor, Frederick W., 14, 158, 179
Team
    investigations, 135–39
    responsibilities of management, 135–36
The Team Handbook, 160
Teamwork, 185–86
Technical Assistance LTA, 84
Technical information systems LTA, 56–60, 91
Technocracy, 14
Technology Solutions Co., 167
Test and qualification LTA, 97
Theodore, Chris A., 20
Theory of knowledge, flow diagram of the, 170
Theory of scientific management, 14
Total Quality Management, 169–95. *See also:* Management; Process improvement; Safety management; *and* System of profound knowledge (Deming)
    constancy of purpose required to improve process under, 174–76
    Deming's Fourteen Points of, 172–94
    mandate of, 2
    quality objectives of, 2–3
    removing impediments to achieving, 177–86
    use of statistical methods in, 26–27
Toynbee, Arthur, 9
Training, 187–89. *See also:* Personal performance discrepancy (error)
    and help LTA, 64
    LTA, 73, 75, 84
    motivation, 99
    personnel, and qualification LTA, 99
Transport LTA, 80–81
Traveler's Insurance Company, 14
Tribus, Myron, 2, 17, 145, 179
Turk, William, 165
Tveite, Michael D., classification of Deming's Fourteen Points by, 173–74

–U–

United States
    Atomic Energy Commission (AEC), 27
    Department of Commerce, Bureau of the Census, 20

–V–

Variation(s)
    cardinal principles of, 23
    knowledge about, 170
    measuring the characteristics of, in safety system performance, 152
    normal versus special cause, 23
    Principles of System, 19–24, 151
Vigor and Example LTA, 90
Vital Few, problems classified as, 137–38, 166

–W–

Walton, Mary, 127
Waste(s)
    elimination of, through quality management, 165–66
    MORT delineated, 50
Work
    environment LTA, 97
    standards (quotas), elimination of, 177
Worker's Compensation, 13–14
    Act (1908), 10
    theory of, 11

–Z–

Zero defects, 192–93